TOWARDS SYSTEM SAFE

Springer
London
Berlin
Heidelberg
New York
Barcelona
Hong Kong
Milan
Paris
Santa Clara
Singapore
Tokyo

Related titles:

Industrial Perspectives of Safety Critical Systems
Proceedings of the Sixth Safety-critical Systems Symposium, Birmingham 1998
Redmill and Anderson (eds)
3-540-76189-6

Safer Systems
Proceedings of the Fifth Safety-critical Systems Symposium, Brighton 1997
Redmill and Anderson (eds)
3-540-76134-9

SAFECOMP '96
Proceedings of the 15th International Conference on Computer Safety, Reliability and Security, Vienna, Austria 1996
Schoitsch (ed.)
3-540-76070-9

SAFECOMP '97
Proceedings of the 16th International Conference on Computer Safety, Reliability and Security, York, UK 1997
Daniel (ed.)
3-540-76191-8

Felix Redmill and Tom Anderson (Eds)

Towards System Safety

Proceedings of the Seventh Safety-critical Systems Symposium, Huntingdon, UK 1999

Safety-Critical
Systems Club

Springer

Felix Redmill
Redmill Consultancy
22 Onslow Gardens
London N10 3JU, UK

Tom Anderson
Centre for Software Reliability
University of Newcastle-upon-Tyne
Newcastle-upon-Tyne NE1 7RU, UK

ISBN 1-85233-064-3 Springer-Verlag Berlin Heidelberg New York

British Library Cataloguing in Publication Data
Towards system safety : proceedings of the seventh
safety-critical systems symposium, Huntingdon, UK 1999
1.System safety - Congresses 2.Industrial safety -
Congresses
I.Redmill, Felix, 1944- II.Anderson, T. (Thomas), 1947-
620.8'6'0285
ISBN 1852330643

Library of Congress Cataloging-in-Publication Data
Safety-Critical Systems Symposium (7th : 1999 : Huntingdon, England)
Towards system safety : proceedings of the seventh Safety-Critical
Systems Symposium, Huntingdon, 1999 / Felix Redmill and Tom
Anderson, eds.
p. cm.
Includes bibliographical references (p.).
ISBN 1-85233-064-3 (alk. paper)
1. Industrial safety--Congresses. 2. Automatic control-
-Reliability--Congresses. 3. Computer software--Reliability-
-Congresses. I. Redmill, Felix. II. Anderson, Tom, 1947-
III. Title.
T55.A1S225 1999 98-50754
620.8'6--dc21 CIP

Printed in Great Britain

Typesetting: Camera ready by contributors
Printed and bound at the Athenæum Press Ltd., Gateshead, Tyne and Wear
34/3830-543210 Printed on acid-free paper

PREFACE

Each year the Safety-critical Systems Symposium brings together practitioners and researchers in a quest to inculcate a higher degree of safety engineering into the development and operation of critical software-based systems. On this, the Symposium's seventh occasion, it explores recent work and experience which lead us further 'towards system safety'.

This book of the Proceedings covers the entire event. The first paper is the course text of a tutorial run on the first day of the Symposium, included here to provide readers with a coverage of the entire event. The next fourteen papers were presented, on the second and third days, in six sessions: *Safety Cases, Systems Engineering, Safety Analysis and Safety Integrity, Tools for Software Safety, Solving Safety Problems,* and *Questions and Competences.*

Eight of the fourteen papers were authored in industry, four in universities, and two in other research establishments. Four of them report on work outside the UK: in France, Germany, Norway and Brazil.

There are three papers on safety cases, each taking a different perspective. Skogstad from Norway and Boyce and Hamilton of GEC-Marconi both report on experience in the field, the former in attempting to apply European norms to project documentation and the latter in attempting to build up a retrospective safety case. The third paper, by Goodman, takes a more philosophical stance, examining the lack of useful measurement in safety assurance.

Both papers on 'systems engineering', originate from the Defence Research Agency where a great deal of work is being carried out on the systems approach. On the subject of 'safety analysis and safety integrity', first Camargo and de Almeida of Brazil and then Hamilton and Rees provide insights into subjects which are not only topical but also very important. The former authors report on the way in which risk and safety analyses were carried out on a metro system, and the latter attempt to make sense of safety integrity levels which, to many, are puzzling.

Two tool applications are reported on, one from France and the other from Germany, but both falling within the SACRES tool set. Then, two eminently practical papers on current problems, the safety aspects of the year 2000 problem and the partitioning of software in safety-critical systems, are offered, the first by Foord and Smith and the second by Simpson and Ainsworth.

In the final topic, 'questions and competences', first Robertson provides a link back to safety cases with his enquiry into the derivation of confidence in safe design, and then Nolan argues that higher confidence could in fact be achieved by less independence in testing. In conclusion, Malcolm reports on the results of an HSE-sponsored project to propose the 'competences' which should be possessed by personnel working in the safety-critical systems domain.

Such coverage of recent steps towards system safety has only been possible because of the willingness of the authors to prepare and present their papers, and, on behalf of both the delegates at the Symposium and the readers of this book, we thank them for this. And, for her dedication and contribution to her seventh Symposium, we thank Joan Atkinson.

FR and TA
October 1998

CONTENTS

The Safety-Critical Systems Club

sponsor and organiser
of the

Safety-critical Systems Symposium

The Safety-critical Systems Symposium '99 is the seventh in the series of annual symposia organised by the Safety-Critical Systems Club.

The Club was inaugurated in 1991 under the sponsorship of the Department of Trade and Industry and the Engineering and Physical Sciences Research Council (EPSRC) and is organised by the Centre for Software Reliability (CSR) at the University of Newcastle upon Tyne. It is a non-profit organisation, set up to provide a service to all, and it sets out to cooperate with all bodies involved or interested in the development or operation of safety-critical systems. Its purpose is to raise awareness of safety issues and to facilitate technology transfer and collaboration among academics and industrialists.

As well as the DTI and EPSRC, the Club is supported by the Health and Safety Executive, the Institution of Electrical Engineers, and the British Computer Society. It seeks to involve both technical and managerial personnel from all sectors of the safety-critical community. By doing so, it facilitates communication between and among researchers and users. It provides a meeting point for industry and academia and a forum for reporting the results of their collaborative projects. The goals of the Club's activities are more effective research, a more rapid and effective transfer and use of technology, the identification of best practice, the definition of requirements for education and training, and the dissemination of information.

Not only does the Club run the annual Safety-critical Systems Symposium each February; it also presents a number of 1- and 2-day seminars and tutorials each year, and publishes a newsletter three times annually.

Membership is open. Since the period of sponsorship expired it has been necessary to request an annual subscription — in order to cover planning, mailing, event organisation and other infrastructure costs — and this can be paid at the first meeting attended. Members pay a reduced fee for attendance at Club events and receive, free of charge, all issues of the newsletter, information on forthcoming events, and any special-interest mail shots. By participating in club activities, they also help to determine the Club's direction.

To join or enquire about the Club or its activities, please contact Mrs Joan Atkinson at: CSR, Bedson Building, University of Newcastle upon Tyne, NE1 7RU; Telephone: 0191 221 2222; Fax: 0191 222 7995; Email: csr@newcastle.ac.uk

Design for Safety

Neil Storey
University of Warwick,
Coventry, UK

1 Introduction

Perhaps an appropriate starting point for a paper entitled 'Design for Safety' is to define what we mean by 'design' and to see how considerations of 'safety' are likely to affect this task. According to the STARTS Guide [STARTS 1987] the design process may be divided into four distinct activities:

- **abstraction**: the operation of generalising, of identifying the essentials;
- **decomposition**: the process of reducing an object into a number of simpler, smaller parts; analysis of interactions, interfaces and structures; modularization;
- **elaboration**: the operation of detailing, adding features;
- **decision making**: identification and selection of alternative strategies.

The goals of the design process are usually manifold. Clearly the resulting system must satisfy its functional requirements, and will normally also have to fulfil certain non-functional requirements which might include such factors as: size; weight; cost and power consumption.

When the operation of a system has implications for safety, the system will also have a set of *safety requirements*. These will define what the system must and must not do in order to ensure the safety of the system. These safety requirements sit alongside the functional requirements to define what the system developer must achieve.

The ability of a system to satisfy both its functional and its safety requirements is limited by the presence of *faults* within the system. Although definitions of this term vary, here the term 'fault' is taken to mean any kind of defect within the system. Faults may be

- **random**: such as hardware component failures
- **systematic**: such as software or other design faults.

Random faults may be investigated statistically and given appropriate data it may be possible to make predictions concerning the probability of a component failing within a given period of time. Systematic faults are not random and are thus not susceptible to statistical analysis. It is therefore much more difficult to predict their effect on the reliability of a system.

Faults may be classified in a number of ways. For example they may be subdivided by considering their nature, their duration or the amount of the system that they affect (their extent). When considering design issues it is convenient to look at three distinct forms of fault, these are:

- random hardware component failures;
- systematic faults in the design (including both hardware and software);
- errors in the specification of the system.

The task of producing a dependable (or safe) system, can be viewed as the process of coping with these various forms of fault in an efficient manner. This can be seen as a process of *fault management* [Storey 1996] and appropriate design techniques can play a crucial part in this process. Fault management techniques may be divided into four groups of techniques:

- fault avoidance
- fault removal
- fault detection
- fault tolerance

Fault avoidance techniques aim to prevent faults from entering the system during the design stage. The avoidance of faults is the primary aim of the entire design process and we shall look at several aspects of this task in this paper. **Fault removal** attempts to find faults within a system before it enters service. Such activities often come under the heading of 'verification and validation'. **Fault detection** techniques are used during service to detect faults within the operational system so that their effects may be minimised. These may include both hardware and software techniques. **Fault tolerance** aims to allow a system to operate correctly in the presence of faults and again, both hardware and software methods may be used. None of these approaches offers a total solution to the problems of fault management and so most critical projects will incorporate a combination of techniques.

When considering the design aspects of a project we can identify three approaches to fault management that are of particular importance:

System Architecture The system architecture has an enormous effect on the ability of a system to tolerate faults within it. It can provide some protection against random component failure and some forms of systematic fault. It does not usually tackle the problems associated with specification faults.

Reliability Engineering This is primarily concerned with the susceptibility of a system to random hardware component failures. However, some engineers believe that these techniques may also be applied to some systematic faults.

Quality Management Considerations of quality cover all aspects of a system's life and are therefore of great importance to fault management.

1.1 Design within the Development Lifecycle

Typical models of the development lifecycle often compartmentalise design into a small number of distinct phases. This is illustrated in Figure 1 which shows the well known 'V' lifecycle model.

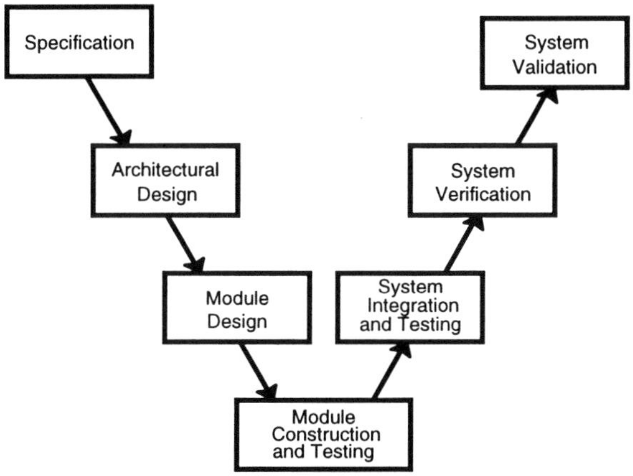

Figure 1 A typical 'V' lifecycle model.

In practice, design activities are carried out throughout the development lifecycle. In order to produce a cost estimate for the development of a system some rudimentary design must be performed at a very early stage of the work, often as part of a feasibility study. The main design work takes place within the 'top-level' and 'detailed' design phases, but later stages of the work will often have a significant design component to produce modifications and improvements. Even during the maintenance phase design modifications may be needed for system upgrading and to remove 'bugs'. Thus design performs an important role in ensuring safety throughout the life of a product or system.

2 Techniques for Achieving Dependability

While there are many techniques for improving the dependability of a system, here we will concentrate on two techniques of great importance to system design. These are the use of fault tolerance and reliability engineering.

2.1 Fault Tolerant Techniques

All real systems are susceptible to faults. The goal of fault tolerance is to design a system in such a way that these faults do not result in a system failure. All methods of fault tolerance are based on some form of **redundancy**. This involves having more complexity within a system than would be required in the absence of faults. Most early forms of fault tolerance used additional hardware to provide protection against hardware component failures. More recently, many forms of redundancy are used to provide protection against a wide variety of faults.

The most widely used forms of redundancy are:

- Hardware redundancy
- Software redundancy
- Information redundancy
- Temporal redundancy

Practical fault tolerant systems use a judicious mix of techniques to provide protection against a range of possible faults.

A simple example of hardware redundancy is the triple modular redundancy (TMR) arrangement of Figure 2. This uses three identical modules which each receive the same input signals. If all the modules are functioning correctly they will all produce identical outputs, and any difference between their outputs will therefore indicate the failure of one of the modules. A voting arrangement is used to remove the effects of any single failure by taking the majority view in the case of disagreement and producing this at its output. The system can therefore tolerate the failure of any single module without this affecting the system's output.

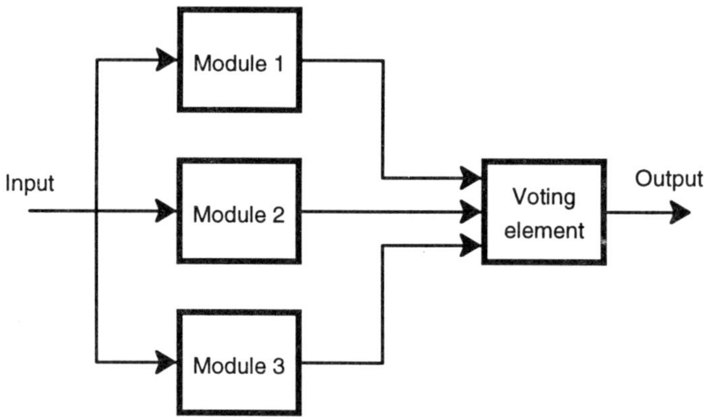

Figure 2 A Triple Modular Redundancy (TMR) arrangement.

The TMR arrangement illustrates the basic concept of redundancy in that three modules are used in a system where a single module could provide the required functionality. It can also be used to illustrate one of the potential problems associated with all forms of redundancy - that of common-mode faults. The TMR approach is based on an assumption that faults are likely to affect the various modules *independently*. This technique provides some protection against random hardware component failures but not against any systematic faults. Since the three modules are identical any systematic fault (such as a software fault) will affect each module in the same way. In this situation the outputs of the modules may be identical but incorrect, and the redundancy achieves nothing. Failures as a result of similar faults in each module are referred to as **common-mode failures**.

Thus TMR provides protection against random component failures, but not against systematic design faults or specification errors.

Common-mode problems are normally tackled by the use of **design diversity** where the redundant modules have the same functionality, but are designed by different teams to try to prevent common faults as a result of common design mistakes. This tackles the problem of common-mode failure at the cost of increased development effort. Redundancy with design diversity can provide protection against both random and some systematic faults, but generally does not tackle the problems of errors in the system's specification, since the same specification will normally be used for each of the diverse implementations. Also, it should be noted that design diversity does not completely eliminate common-mode design errors since it has been demonstrated that different design teams are likely to make similar errors.

2.1.1 Fault Detection Techniques

Many fault tolerant techniques rely on the use of some form of fault detection mechanism. These look for errors produced during the operation of the system. Examples of fault detection methods include:

- Functionality checking
- Consistency checking
- Signal comparison
- Checking pairs
- Information redundancy
- Instruction monitoring
- Loopback testing
- Watchdog timers
- Bus monitoring
- Power supply monitoring

It is not within the scope of this paper to describe each of these techniques in detail. An overview of these methods can be found in [Storey 1996].

2.1.2 Hardware Fault Tolerance Techniques

Hardware fault tolerance can be achieved using three basic techniques:

- Static redundancy
  Uses some form of voting network as in the TMR arrangement
- Dynamic redundancy
  Relies on some form of fault detection and reconfiguration
- Hybrid redundancy
  Uses a combination of the above techniques

Static Redundancy

Static redundancy uses a number of redundant modules and some form of voting arrangement as in the triple modular arrangement described earlier. Use of three modules in a TMR scheme provides a system that will work correctly if a single module fails, and because of the effects of the voting network, the effects of the failure are *masked* from the outside world. If more redundant modules are used it is possible to produce a system that can tolerate a greater number of module failures. Such systems are given the generic name of *N*-modular redundant (NMR) systems. For example an NMR arrangement with 5 modules can tolerate the failure of two modules.

A potential problem with static arrangements is the possibility of a single-point failure within the voter. If the inputs come from a single sensor this also represents a possible cause of single-point failure. Fortunately, the voter is often a very simple arrangement and it is possible that this may be sufficiently dependable to produce an acceptable performance. Similarly, some simple sensors may have a very high dependability. However, in critical applications it is more common to use multiple sensors and multiple voters as shown in Figure 3. Such an arrangement will tolerate the failure of a single module, a single sensor or a single voter. The multiple outputs may be fed to later fault tolerant stages to form a multiple stage TMR arrangement, or may be fed to multiple output actuators that effectively vote at the output.

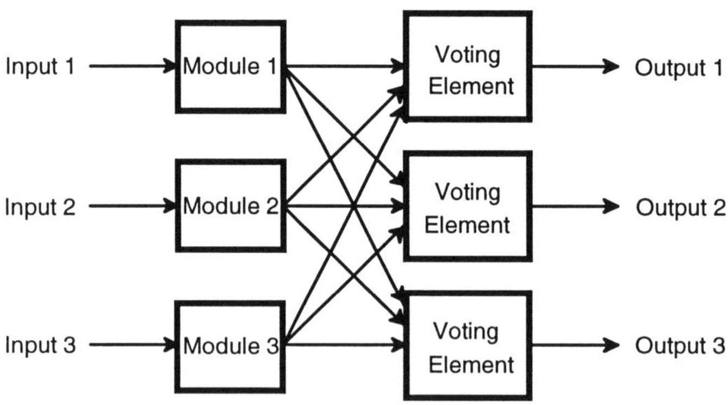

Figure 3 A TMR arrangement with multiple inputs and voters.

Dynamic Redundancy

Dynamic redundancy uses fault detection and reconfiguration to achieve fault tolerance rather than voting. This results in an arrangement that requires less redundant hardware to achieve the same degree of fault tolerance, but places great reliance on the fault detection methods used. Also, in many cases, once a fault has been detected it will take a finite amount of time to reconfigure the system to remove the effects of the faulty unit. During this time the output of the system may be incorrect. Dynamic redundancy does not provide fault masking.

A simple example of dynamic redundancy is the standby spare arrangement shown in Figure 4. Here one module is operated in conjunction with some form of fault detection mechanism. While no fault is detected the output from this module is fed to the system's output by a switch network. If a fault is detected the detection mechanism will reconfigure the system so that the switch takes its output from the standby module.

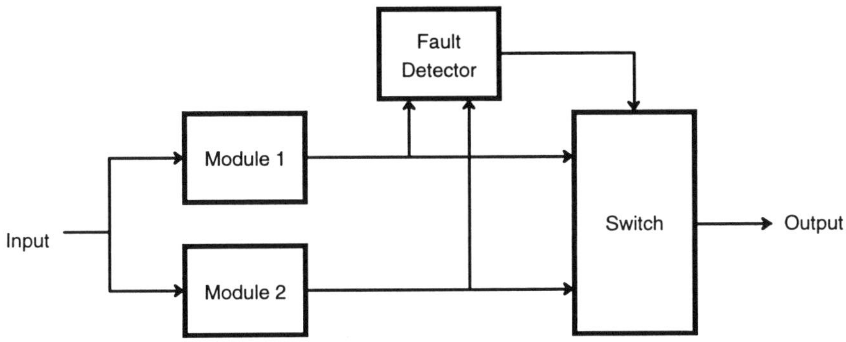

Figure 4 A standby spare arrangement

The process of reconfiguration will cause a temporary disruption of the output while the output is switched. This disruption can be minimised by the use of a *hot standby* where the backup module is running continuously in parallel with the active module. This permits the operation to be switched very quickly to the standby unit. Unfortunately a disadvantage of this arrangement is that the standby module is subject to the same operating stress as the active unit and so will wear in a similar manner. It will also consume power at the same rate. An alternative strategy uses a *cold standby* arrangement where the backup module is unpowered while not it use. This reduces power consumption and wear in the unit but means that the transition to the alternative module may take much longer in the event of a failure of the active unit.

The standby spare arrangement permits a single module failure to be tolerated using a total of two modules, rather than the three required using TMR, and adding extra redundant modules allows more failures to be tolerated. For example, a dynamic arrangement using three modules (an active module plus two spares) permits two module failures to be tolerated. Thus dynamic redundancy offers advantages in respect of the amount of redundant hardware required. However, the success of the arrangement relies on the effectiveness of the fault detection hardware.

Another example of dynamic redundancy is the self-checking pair arrangement. Here two identical modules are fed with the same input signals and their outputs are compared as a form of fault detection. The output from one of the modules is passed to the next stage and the output of the comparator is used as a failure detection signal. The signal comparison may be performed in hardware or in software. The self-checking pair does not in itself produce fault tolerance, but does provide a block with built-in fault detection that can be used in a dynamic fault detection arrangement. A simple self-checking pair is shown of Figure 5.

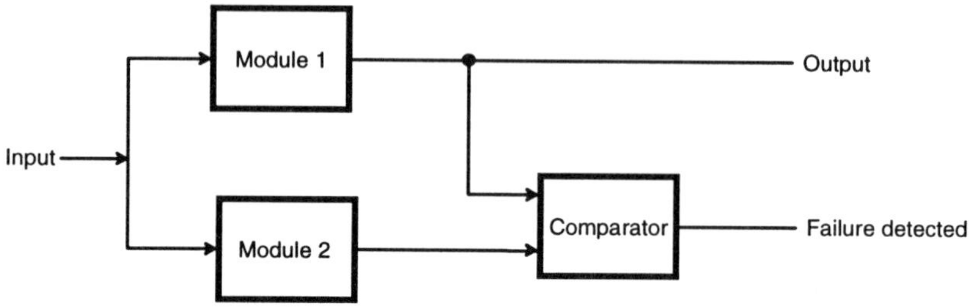

Figure 5 A self-checking pair

Hybrid redundancy

We have seen that static redundancy provides fault masking but requires more redundant hardware than dynamic arrangements. Hybrid redundancy uses a combination of voting, fault detection and reconfiguration to achieve some of the advantages of both techniques. There are many forms of hybrid redundancy but most can be generalised to some form of N-modular redundancy with spares as shown in Figure 6.

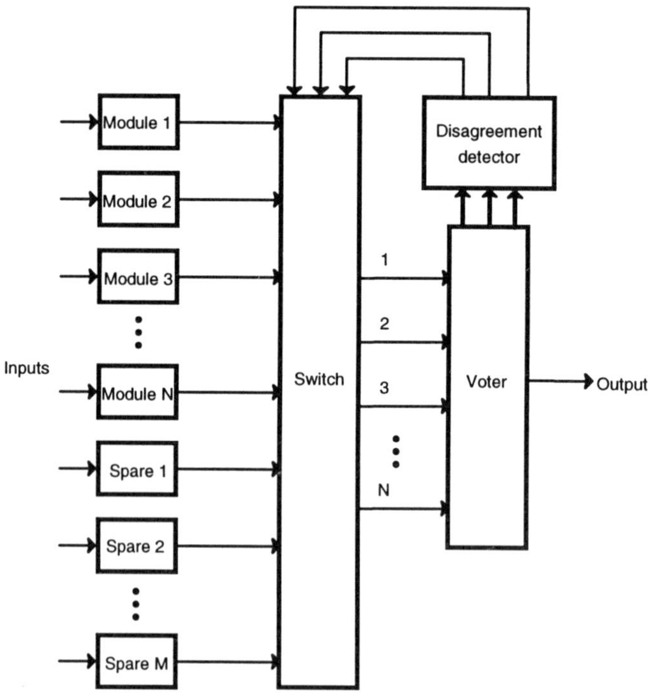

Figure 6 N-modular redundancy with spares

2.1.3 *Software Fault Tolerance Techniques*

Fault tolerance can be achieved not only through the use of redundant hardware, but also through the use of appropriate software techniques. Many hardware fault tolerant arrangements make use of a number of identical modules. As discussed earlier this provides protection against random component failures within the modules but does not tackle the problems associated with systematic design faults. Software faults are *always* systematic, so the simple duplication of software does not provide any protection against such faults. In order to achieve fault tolerance we must use diverse software.

There are two common methods of achieving fault tolerance within software, these are:

- *N*-version programming
- Recovery blocks

Unfortunately, there is not scope within this paper to describe these techniques in detail. However, it is worth noting that both approaches make use of design diversity and that *N*-version programming [Avizienis 1985] can be thought of as a software equivalent of the static hardware techniques discussed earlier, while recovery blocks [Anderson 1990] are similar in many ways to dynamic methods of hardware fault tolerance.

2.2 Reliability Engineering

Components that fail as a result of non-systematic faults will fail at a random time. For a given device it is not possible to predict when failure will occur, but it is possible to quantify the rate at which members of a family of components will fail.

Systematic faults are not random and therefore are not subject to statistical analysis. We have also seen that fault tolerant techniques based on the use of identical hardware modules provide no protection against systematic faults. For these reasons systematic faults represent a very serious problem for the system designer. In computer-based systems one of the most common forms of systematic faults is the software 'bug'.

Reliability engineers are divided on how to approach the problems of software faults. Some say that since they occur each time a given piece of code is executed they are totally predictable and so are not susceptible to statistical analysis. This leads to the view that we cannot apply a figure to the reliability of software. An alternative view is held by engineers who feel that it *is* appropriate to use statistical techniques with software. They argue that because of the complexity of software, faults could take an almost unlimited number of forms and that consequently their distribution can be considered to be random. This leads to the view that *unknown* software faults are sufficiently random to allow statistical techniques to be applied and hence to the belief that reliability engineering *is* of relevance to software. For the moment we will ignore the issue of the applicability of these techniques and look instead at the analytical methods themselves.

When reliability is treated quantitatively it is normal to define it as the probability of a component or system working correctly over a given period of time. Since reliability is a function of time it is common to give it the symbol $R(t)$.

Closely linked to considerations of reliability is the failure rate of a component. Experience shows that for a wide range of components, such as electronic devices, the failure rate has a distinct time varying nature as shown in Figure 7. For obvious reasons this characteristic shape is described as a 'bathtub' curve.

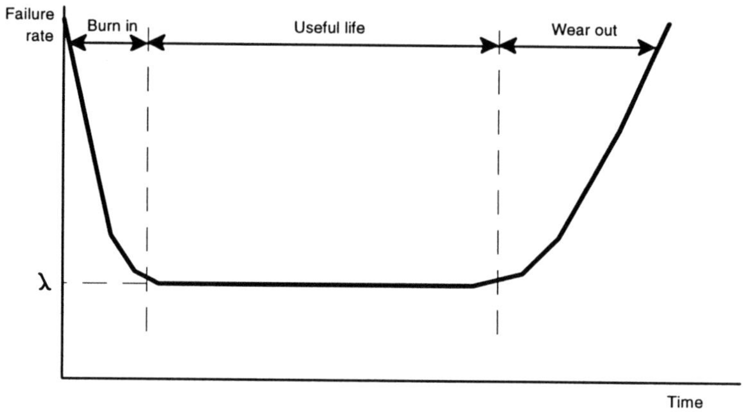

Figure 7 Typical variation of failure rate with time

The curve shows three distinct phases during the component's life. In the early stages the device suffers a high failure rate as a result of 'infant mortality' owing to the presence of manufacturing faults not detected during testing. This is followed by a region of approximately constant failure rate. This is termed the 'useful life' portion of the curve and manufacturers will aim to use the component only during this phase. Towards the end of the component's life the failure rate rises again during what is termed the 'wear out' phase. In critical applications components should be replaced before they enter this stage.

It can be shown that during the constant failure rate stage the reliability of a component is related to its constant failure rate (λ) by the expression

$$R(t) = e^{-\lambda t}$$

This relationship is termed the exponential failure law, and shows that when the failure rate is constant, the reliability falls exponentially with time.

The failure rate of some components is not constant but varies with time. The failure of mechanical components due to fatigue and failures due to software bugs have time variant characteristics. This form of behaviour may be described by the Wiebull distribution which is of the form:

$$R(t) = e^{-(t/\eta)^{\beta}}$$

While time-variant failure rates are of great importance, here we will restrict ourselves to looking at components that exhibit a constant failure rate.

2.2.1 *Reliability Modelling*

A common problem in reliability engineering is to predict the reliability of a complex system from a knowledge of the failure characteristics of its components. One way of tackling this problem is by the use of 'combinational modelling'. Many text books give a good introduction to this technique (for example [Lewis 1996]) and here we will look at some of the results that may be obtained using this approach rather than consider the technique itself.

One of the results that may be obtained using combination modelling is an estimate of the overall reliability of a TMR arrangement $R_{TMR}(t)$ in terms of the reliability of the individual modules $R_m(t)$. This relationship is

$$R_{\text{TMR}}(t) = 3R_m^{\ 2}(t) - 2R_m^{\ 3}(t)$$

This equation ignores the effects of the voter although combinational modelling can be used to describe the effects of voter failure on reliability.

To illustrate the characteristics of this expression let us look at some sample values. For example, if for a given period of time t the reliability of a module is equal to 0.95, then the expression shows that over this period of time the reliability of a TMR arrangement constructed using three identical modules would be 0.993. It can be seen that the redundant arrangement has created a significant increase in reliability. However, it should not be assumed that this will always be the case. Consider, for example, a TMR arrangement using three modules each with a reliability of 0.4. In this case the reliability of the overall arrangement is now 0.352, which is lower than that of the individual components. In order for a TMR arrangement to give an increase in reliability, the individual modules must have a reliability of greater than 0.5. Reliability cannot be achieved by combining unreliable components.

Another point to note is that reliability is a function of time and that the reliability of all components falls as they are used over longer periods. Consequently the reliability of all components will fall below 0.5 at some point in their life. Therefore at some stage a TMR arrangement will always be less reliable than an implementation using just a single module. This emphasises the importance of a careful choice of system architecture and the need to consider carefully the period of time over which reliability calculations are performed.

Combinational modelling can also be used to estimate the reliability of a system with dynamic redundancy. For example, a hot standby spare arrangement using two identical modules each of reliability $R_m(t)$, has an overall reliability of:

$$R(t) = R_m(t) + [1 - R_m(t)]C_m R_m(t)$$

where C_m is the is the fault coverage of the module and represents the probability of a fault within the module being detected. It can be seen that the reliability of the system is greater than that of an individual module by an amount that is determined by the ability of the system to detect a module fault. The expression for the reliability of a cold standby arrangement is somewhat different from the above equation, but is of the same basic form.

Inherent in the above calculations is the assumption that failures within the various modules are completely independent. This will be the case for random component failures, but not for systematic faults which could affect all the modules simultaneously. We have also assumed that each module has a constant failure rate.

2.2.2 *Reliability Prediction*

Clearly it would be useful to be able to predict the reliability of a system at the design stage. In the last section we saw how we could determine the reliability of a system from the characteristics of its components, but this still leaves us with the problem of predicting the failure rate of its components, and of subsystems that have yet to be built.

The most widely used technique for predicting the failure rate of electronic components is based on the United States Department of Defense military handbook MIL-HDBK-217 [DoD 1992]. This aims to predict the failure rate of components from a knowledge of their form and their operating environment. The handbook provides models for a wide range of components which can be used to calculate the predicted number of failures per million hours of operation from a number of parameters. For example, the model for a simple resistor is:

$$\lambda_p = \lambda_b \pi_R \pi_Q \pi_E \text{ Failures}/10^6 \text{ Hours}$$

where λ_p is the part failure rate, λ_b is the base failure rate reflecting the electrical and thermal stresses on the part and the various π terms represent environmental and other effects. Here π_R models the importance of the resistor value, π_Q concerns the quality of its manufacture and π_E is an environmental factor. Values for the various parameters are given in tables in the handbook.

The values determined using the models within the handbook can be used to provide an indication of the expected reliability of the final system but these should not be seen as providing an accurate estimate. They are intended to be used to guide design choices rather than to provide an accurate estimate of reliability.

We noted earlier that engineers are divided on the applicability of reliability measures to software. Where they are used they often aim to predict the reliability that might be expected from a piece of software. Such techniques are usually based on an estimate of the number of residual errors within a program. This is normally estimated by considering the number of errors that are found during development and attempting to extrapolate to provide an estimate of the number of remaining errors. There are many problems with this approach since there would seem to be no direct correlation between the number of errors within a piece of software and its failure rate. Such factors as the form of the error, its effects and the frequency with which it is executed will also affect failure rate. A great deal of work is in progress in an attempt to allow the reliability of software to be predicted [Littlewood 1993, Pyle 1991]. However, at present there is no universally accepted method of achieving this.

2.2.3 *Reliability Assessment*

Once a system has been designed and a prototype system constructed it is necessary to demonstrate that it meets its requirements before it goes into service. One of these requirements will be its need for a certain level of reliability. Unfortunately, in many cases the level of reliability required of a critical system is beyond our ability to demonstrate by testing alone.

IEC 1508 [IEC 1995] gives a list of target failure rates for systems of different levels of integrity. These are shown in Table 1. It can be seen that some systems require a mean time to failure (MTTF) of 10,000 years or more. To demonstrate this by testing is impossible. It is generally agreed that these levels of failure rate are several orders of magnitude better than can be demonstrated using our current techniques.

Table 1 Target Failure rates from IEC 1508

Safety integrity level	Continuous mode of operation (probability of a dangerous failure per year)	High demand/continuous mode of operation (probability of failure to perform its designed function on demand)
4	$\geq 10^{-5}$ to $< 10^{-4}$	$\geq 10^{-5}$ to $< 10^{-4}$
3	$\geq 10^{-4}$ to $< 10^{-3}$	$\geq 10^{-4}$ to $< 10^{-3}$
2	$\geq 10^{-3}$ to $< 10^{-2}$	$\geq 10^{-3}$ to $< 10^{-2}$
1	$\geq 10^{-2}$ to $< 10^{-1}$	$\geq 10^{-2}$ to $< 10^{-1}$

Since we cannot demonstrate these levels of performance by testing we need to find other ways of convincing ourselves, and possibly a regulator, that our systems meet their needs. Invariably this requires that we adopt certain design and development methods that can give us confidence in the correctness of our system.

3 Design Issues

3.1 Hardware Design Issues

There are a great many issues of importance to the design of safety-critical computer hardware. These include topics that we have already considered such as fault tolerant architectures and reliability engineering, as well as other areas such as real-time system design and EMC. In this brief paper we will look at just a few topics which have been chosen because they have a direct relevance to safety and because they are often ignored in the literature.

3.1.1 *Microprocessor Design Faults*

Modern microprocessors are extremely reliable and in most applications the probability of failure due to an inherent fault within the chip is small compared to the probability of failure due to other causes (such as software faults). For this reason in most commercial applications microprocessor design faults are normally ignored. However, in highly critical applications all possible faults must be considered.

Faults within the design of the processor are systematic and so are not tackled by simple fault tolerant techniques using identical modules. Also, the complexity of modern devices makes the variety of possible faults almost infinite, while making exhaustive testing impossible.

While there are a great number of possible forms of design faults, these may be divided into two main categories:

- Failure of the circuit to correctly implement its intended function
- Failure of the documentation to correctly describe the circuit's operation

It might seem at first sight that only the first of these classes represents what might be termed *design* faults, since the second is simply a weakness in the documentation. However, as far as the user of a processor is concerned, the manual represents the 'definition' of the device, and any discrepancy between this and the actual operation has exactly the same effect as a functional error.

While there is little documentary evidence on the statistics of microprocessor faults it is clear that these are most likely to be associated with instructions that see limited use or with events that occur infrequently. This is because such faults are likely to go undetected during testing. In one of the few papers on this topic Wichmann reported that a particular processor's operations were completely incorrect for certain instructions that were not generated by the associated C compiler [Wichmann 1993]. This was no doubt because the manufacturer's testing procedures involved the execution of large numbers of test programs which were all written in C.

Experience shows that early versions of a device have significantly more problems than later releases. This is because manufacturers progressively remove faults located in service. However, the fact that processors are continually evolving does itself pose a problem for the system developer, since each new version will have slightly different characteristics. This can invalidate testing performed using earlier parts.

Because major design errors would be discovered during system development, it follows that most microprocessor faults are relatively subtle. For this reason they often go undetected when the device is incorporated into a new application. Well known examples of microprocessor faults include the indirect jump instruction in the 6502 processor which operates incorrectly if the indirect address happens to straddle a page boundary, and the widely publicised divide problem in the Pentium processor which related to a fault in a rarely executed arm of a case statement.

In many cases processor design faults are quickly located and design teams who have experience with a particular device find methods of working around the problem. However, the faults are rarely documented and so this information is not distributed to other users. In critical applications this situation is very dangerous.

3.1.2 Choosing a microprocessor

The choice of a processor for any real-time application will involve a large number of factors. These will include the processing power required and any particular architectural requirements. In safety-related applications there are also considerations related to the integrity requirements of the system. From the last section it is clear that the presence of known design faults within a processor is likely to affect our choice but there are also other issues of importance.

Certain processors have characteristics that, though not design faults as such, make them unsuitable for use in critical applications. Though obsolete now, a good example of such a device is the Motorola 6801 processor. This has within its instruction set a test instruction that fetches an infinite number of bytes from memory. This produces a regular pattern on the address bus which is extremely useful during testing. This instruction is not intended for normal use and does not appear in the list of instructions given in the manual. However, if this instruction is executed inadvertently (perhaps as a result of a noise-induced jump error) it represents a 'black-hole' that is impervious to both maskable and non-maskable interrupts. Recovery from this situation can only be achieved by resetting the device. Clearly a watchdog timer connected to the reset line would protect the system from such an event. However, it should be remembered that this instruction is undocumented, and a designer could perhaps be forgiven for connecting the watchdog timer to the non-maskable interrupt line to allow a unique handler to be used.

Since details of design faults and other unacceptable features are rarely documented, how does a designer choose a processor for a given critical application? In some cases there will be experience within a company or organisation that can guide this choice. If not, then engineers should be guided by the choices being made by other companies. While it is unlikely that competing companies will publicise the components they are using, one can tell which components are being widely used by looking at the support tools available. Developing safety-critical systems requires some very specialist tools - particularly in the area of software testing. These tools are generally specific to one microprocessor and are very expensive to develop. Consequently they are only available for a small number of components that are widely used in critical systems. Thus looking at tool availability provides a good guide to component suitability for critical applications in general. It does not however, show that a part is ideal for your particular application.

In general the devices that are widely used in critical applications are not 'state-of-the-art' components. Designers usually prefer to use well tried and tested components rather than the newest parts. In such applications a 'track record' is often far more important than pure processing power.

3.2 Software Design Issues

In computer-based systems much of the complexity is implemented within the software and for this reason many of the associated problems are software-related. Some engineers see the task of developing a critical system as largely the problem of getting the software right. Unfortunately getting software to work correctly is not easy and even simple programs usually do not work first time. In non-critical applications the task of software generation is generally an iterative process of writing, testing and modifying the code until it appears to work. The development process ends when the software stops failing the series of tasks being used to test it. Thus the 'dependability' of the software is determined almost exclusively by how it is tested.

Unfortunately testing is notoriously difficult and exhaustive testing is almost always impossible. When developing a critical system this kind of 'dynamic' testing is supplemented by various forms of 'static' testing which look at the characteristics of the code without executing it.

While the testing of safety-critical software is not within the scope of this paper, considerations of how a system is to be tested have many implications for the way it is designed. Here we will look at one such issue - the choice of programming language to be used.

3.2.1 Choice of Programming Language

In commercial applications important considerations in the choice of a programming language include such factors as productivity, efficiency and portability. For this reason, most commercial software is written in languages such as C, C++ and Java.

In critical applications the above considerations are also going to be of important, but other factors will also need to be considered. These include the characteristics of the language and the availability of support tools.

Language Characteristics

Programming languages differ wildly in their appropriateness for use in safety-related systems. Carré *et al.* identified six factors that influence the suitability of a language for high-integrity applications [Carré 1990]. These are:

- Logical soundness
- Complexity of definition
- Expressive power
- Security
- Verifiability
- Bounded time and space constraints

No standard programming language performs well in all these areas although some (such as Pascal and Ada) perform much better than languages such as C or C++. In highly critical applications 'verifiability' is of great importance. Certain languages allow powerful software verification tools to be used to perform a wide range of static tests on the code to detect a range of programming errors.

Tools support

An important issue in the selection of a programming language is the quality of the available compilers and other tools. For certain languages *validated* compilers are available. While not guaranteeing perfection, validation greatly increasing our confidence in a tool. Unfortunately, validated compilers are only available for a limited number of languages, such as Ada and Pascal.

In addition to compilers, developers of critical systems will make use of a range of other tools such as static code analysis packages. The static tests that can be performed on a piece of code vary greatly depending on the language used. To aid this process it is common to restrict the features that are used within certain languages to a 'safe subset' of the language. Well structured and defined languages such as subsets of Ada, Pascal and Modula-2 allow a great many tests to be performed such as data flow analysis, data use analysis, information flow analysis and range checking. Unfortunately many of these tests cannot be performed on languages such as C and C++ .

Choosing a language for a given application

Invariably several factors will determine the language that is chosen for a given project. In some cases the language to be used may be defined by the 'customer'. For example, much of the work performed for the US Department of Defense must be performed in Ada. The various generic and industry specific safety standards also include requirements or guidance on the choice of language. The languages that are preferred within these documents vary between industries and with the level of integrity required. Projects that require a very high level of integrity often use safe subsets of languages such as Ada and Pascal. Systems with a lower safety requirement often use full versions of Ada, ISO Pascal, Modula-2 or structured assembly code.

Perhaps one of the most contentious issues within this field is the use of C and C++ in safety-related applications. Some practitioners feel that these languages should not be used at all in such systems because of their poor definition and lack of structure. However, in some areas, such as the automotive industry, the use of C is widespread even in safety critical applications. To try to reduce some of the problems associated with the use of C in such situations, and to promote the use of agreed 'best practice', the motor industry has recently produced a new set of guidelines in this area [MISRA 1998].

3.3 Human Factors

When designing safety-critical systems we strive to make them as simple as possible. Unfortunately, many systems have within them a very complex component - a human operator. Humans have the advantages a flexibility and adaptability but in many situations they are unreliable and unpredictable.

Many experiments have been performed to measure the error rates associated with human operators performing a range of task. While the data produced by these tests is very dependent on a wide range of circumstances, it is clear that the error rates observed are much higher than one would expect from any form of automated system. Typical values for an operator performing very simple tasks might be one error in every 1,000 to 10,000 operations. While for more complex tasks this error rate might increase to one error in every 10 or less operations.

In order to remove the likelihood of an accident being caused by human error it is normal to attempt to remove the human operator from all responsibility for safety. If an operator is needed, he or she may be given a supervisory role with the required safety features being implemented within an automated system. While this may reduce the probability of an accident being caused by the operator making a mistake, it does not remove the problems of human error or reduce the importance of human factors. Automated systems are designed, manufactured, installed and maintained by humans, and errors by any of those involved can influence the system's safety. While in some cases you may be able to remove the responsibility for safety from an operator, you generally cannot remove it from the system developer. Thus great reliance is placed on the development strategies and quality management structures used.

4 The Design Process

We noted earlier that design plays a part in each phase of the development lifecycle. Here we will look at just a few elements of the design process.

4.1 Top Level or Architectural Design

The top-level architectural design phase allocates the various functional requirements of the system to appropriate implementation structures. In safety-related applications it is also necessary to allocate the various safety requirements, identified in early phases of the development, to appropriate safety-related systems or subsystems. In general these will include systems based on a number of technologies and may include mechanical, hydraulic or electrical subsystems, as well as both programmable and non-programmable electronic sections. Wherever possible safety features should be implemented using the simplest possible elements.

Top level design also involves partitioning the system functions into those to be produced within hardware and those that are to be implemented in software. When this has been done, the architecture of the hardware and software can be defined. One aspect of this process is the *decomposition* of the system into manageable modules. This involves the specification of the functions and safety features to be implemented by each module and the definition of the interfaces between them. It also involves the definition of the major data structures within the software.

One approach to system design is based on the principle of *hierarchical design*. This technique, which can be applied to both hardware and software, divides the system into a number of layers. Modules within each layer depend for their correct operation on modules in lower levels of the structure. These in turn are dependent on modules within lower levels.

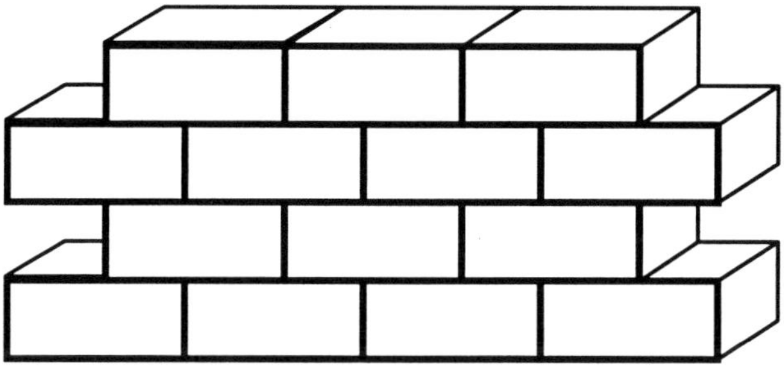

Figure 8 A layered approach to design

This arrangement can be represented diagrammatically as shown in Figure 8. One of the principles of hierarchical design is some form *of downward-only functional dependence*. This results in layers of *abstraction* that allow the functioning of lower levels to be hidden from those above it.

4.2 System Partitioning for Safety

The way in which a system is partitioned is fundamental to the provision of safety. One of the important aspects of partitioning is that it aids *comprehension* of the system. Large monolithic structures of either hardware or software are difficult to understand and are therefore prone to errors. A well partitioned system is much easier to understand.

A second function of partitioning is that it provides a level of *isolation* between the modules. This can be used to contain faults and is fundamental to the provision of fault tolerance. It may also simplify the task of verification by allowing modules to be considered separately.

Because the operation of one module can affect the operation of another the way in which the system is partitioned is of great importance. Consider the software arrangement of Figure 9.

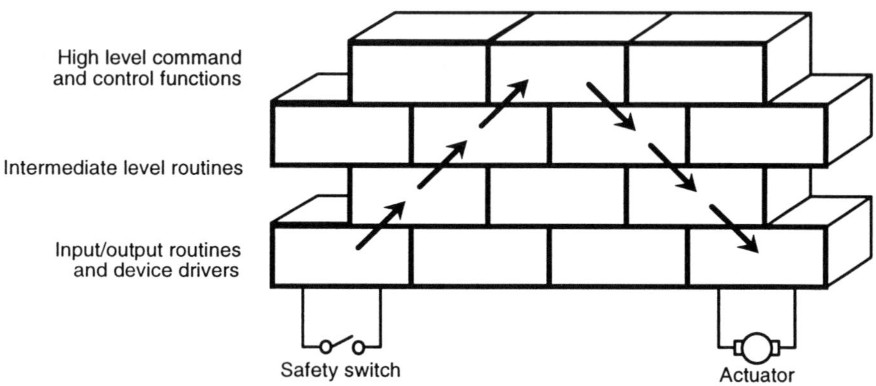

Figure 9 A poorly structured system

In Figure 9 an actuator that is capable of some dangerous effect must be activated by a high-level routine. To do this it calls an intermediate-level routine that in turn calls a lower-level routine, and so on, until the appropriate low-level routine is activated. In order to maintain safety some form of safety switch is used to produce an interlock mechanism. Before the high level routine will operate the actuator it checks the safety switch to see that it is the appropriate state. It does this by calling a series of intermediate-level routines, which in turn call the required low-level routine. Provided that all elements in the system work correctly this arrangement should be safe.

An unfortunate characteristic of the arrangement of Figure 9 is that all the software elements involved are critical to the safety of the system. If any of the high, intermediate or low-level routines malfunctions the actuator can be activated dangerously.

Now consider the arrangement of Figure 10. Here the safety switch and the actuator have been encapsulated within a single low-level module. If the high-level routine now wishes to operate the actuator it sends a message to this low-level routine which is equivalent to 'operate the actuator if it is safe to do so'. Because the low-level routine implements the complete interlock mechanism it is the only critical module within this arrangement. If perfect isolation could be guaranteed between the various modules, only the low-level module would be safety-related. Since the low-level routines are generally also the simplest modules, this would drastically reduce the amount of code that would need to be developed and tested to a high level of integrity. In practice, complete isolation is rarely possible, particularly between software modules, and thus all system elements are likely to be safety-related to some extent. However, the integration of safety functions into simple, self-contained modules that can be extensively tested, is a very attractive design goal. Unfortunately, not all systems lend themselves to this approach and in many cases safety functions require high level control. In such cases much of the system becomes critical to the safety of the system.

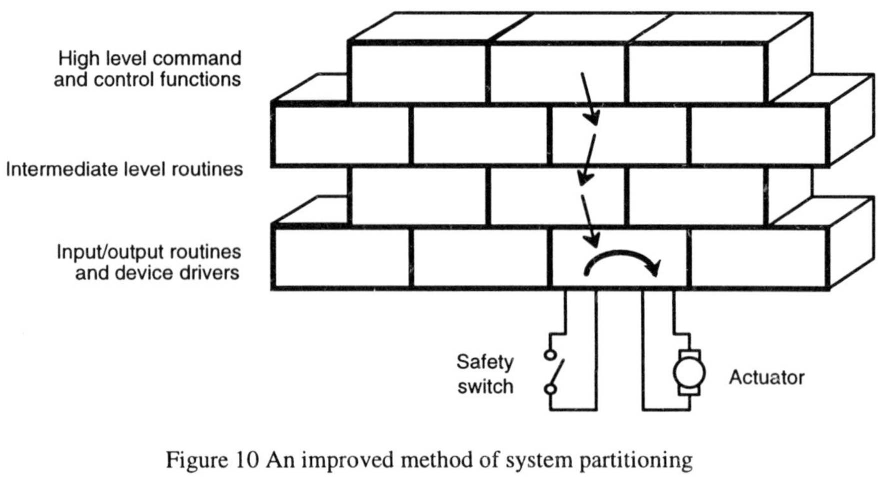

Figure 10 An improved method of system partitioning

4.3 Detailed Design

Following the process of decomposition performed in the top-level design phase comes the detailed design of the various functions of each module. The process of decomposition is often iterative, with modules being broken down successively into small sub-modules, each with its own specification.

The techniques used in the detailed design phase will be greatly affected by the overall development methods and tools being used. For example the use of formal methods may permeate all phases of the design. We shall look at formal methods briefly in Section 5.3. There are also several special-purpose design tools or methods that may be used to facilitate the design process. These include such techniques as Yourdon [Yourdon 1979], Jackson [Jackson 1983] and Mascot [RSRE 1987].

For software, well structured languages such as Ada and Pascal greatly simply the design process as they naturally support an iterative approach to functional decomposition.

4.4 Safety Kernels and Firewalls

In some cases safety can be enhanced by the use of safety kernels or firewalls.

A *safety kernel* consists of a relatively simple arrangement, often a combination of hardware and software. Its small size and lack of complexity enable it to be developed into a trusted subsystem that can be used to ensure the critical safety functions of a system. The success of this arrangement depends on the ability of the designer to protect the kernel from outside influences. This might be achieved through the use of separate hardware, or in the case of a software kernel, through software partitioning.

An alternative approach is to place the components (hardware or software) responsible for safety behind a protective firewall. When considering hardware this could be a physical barrier to protect the critical sections from the dangerous effects of system failure. Within software, the firewall could be a logical barrier providing access control to stop unauthorised access or modification of critical data or code [Leveson 1995].

4.5 Design for Maintainability

Although it may not always be immediately apparent, good maintainability is often a prerequisite of safety. This is true not only because a system that is difficult to maintain is likely to be badly maintained and is therefore likely to be less reliable, but also because maintainability has a direct relationship to availability.

One factor that is often overlooked in the operation of safety-critical systems is the impact of maintenance induced failures. Evidence from a number of sources suggests that there is a significant probability that maintenance operations will not be completed satisfactorily and may lead to new, and seemingly unrelated faults. These problems are likely to be more severe in systems that are difficult to maintain and so careful thought should be given to this aspect of the design.

Linked with this issue is the situation where attempts to increase maintainability may *decrease* the safety of the system. An example might be the inclusion of built-in test equipment (BITE) within a system. BITE speeds maintenance by simplifying the location of faults, but requires an increase in hardware or software and therefore inevitably reduces the overall reliability. There is therefore a trade-off between maintainability and reliability. This needs to be considered carefully at the design stage.

4.6 Component Reuse

The reuse of components of hardware or software makes good economic sense and can also have safety benefits.

In many high integrity projects development costs dominate and the ability to pick up and use a system or subsystem from another project is clearly attractive. Indeed, if a component has a proven history of successful operation in another application, its inclusion would seem to have safety benefits when compared to a untried and untested new component.

When using components from other projects care must be taken to ensure that appropriate steps are taken to make them comply with the requirements of the current system. Full documentation of the development and verification of the component will be required, and these should be reviewed in the light of their changed use. Guidelines on the reuse of components are given in many of the standards used in this area.

While component reuse offers many advantages it can be disastrous if not performed correctly. A memorable example of what can go wrong if this is not done properly is the Ariane V accident in 1996, which was directly attributable to problems associated with the reuse of a system from a previous rocket.

5 Other Considerations

5.1 Commercial Off The Shelf (COTS) Components

An issue closely related to the reuse of sub-systems is the use of 'commercial off the shelf' (COTS) components. This applies to both hardware and software, and in these cost conscious times, COTS components are being increasingly used in a wide range of industries. Programmable logic controllers (PLCs) can be regarded as COTS components and even the five computers within the space shuttle are standard general-purpose computers.

The use of COTS parts can have both advantages and disadvantages. Perceived advantages are: reduced development time and cost; standardisation; increased confidence through component reuse; and the ability to upgrade at a lower cost. Perceived disadvantages are: a difficulty in determining the integrity level of bought in components; limitations in the available documentation; unwanted functionality leading to unwanted complexity; and difficulties in certification because of a lack of basic data.

COTS is the way of the future, although this approach provides a number of challenges, particularly in the area of certification. In some areas there are moves towards standardisation which could ease the situation. For example in Germany the TÜVs are now offering a certification service for PLCs.

5.2 Tools Support

The development of safety-related systems requires the use of a large number of automated tools. Figure 11 shows various classes of tools that might be used within a typical project. The figure differentiates between tools that are used within a small number of phases (vertical tools) and those that are used throughout the development lifecycle (horizontal tools).

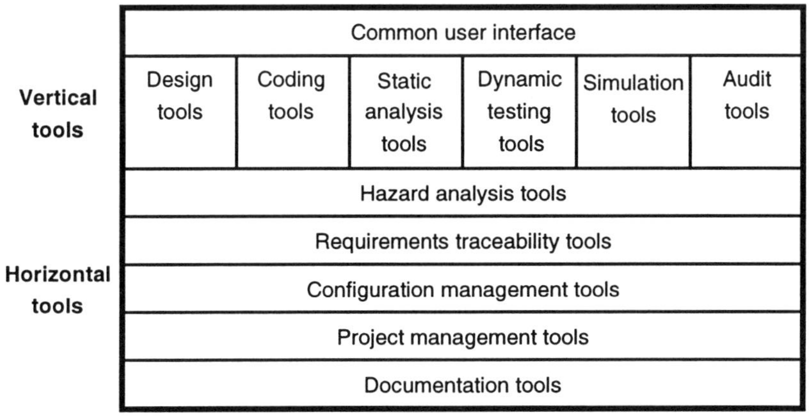

Figure 11 Classes of development tools

Many of the tools associated with the development process have a direct influence on the safety of the resulting system. This is particularly true in the case of tools for static and dynamic testing. For this reason several safety standards give guidance on the selection of tools and the process of 'tool certification'.

If a tool is to be used in the development of a system that is highly critical, then the tool itself must be very dependable or faults within the tool could jeopardise the system's integrity. This leads to the idea of assigning a level of integrity to the development tools as well as to the finished product. Unfortunately, there are many problems associated with this approach, and most tool manufacturers do not claim a particular level of integrity for their products.

The task of selecting and evaluating tools can be both time consuming and expensive. In many cases tools must be compared on the basis of the manufacturer's claims rather than on first hand experience. In assessing such claims the manufacturer's credibility and track record must also be considered.

5.3 Formal Methods of Design

The term 'formal methods' describes the use of mathematical techniques in the specification, design and analysis of computer hardware and software. These methods can be applied to none, some or all of the development phases, and can be applied to part or all of the system.

The design of a system can be viewed as a series of transformations of the system's definition. Initially the system specification describes the system, and this is transformed into a top-level architectural representation which should have the same functionality. This in turn is transformed in the detailed design phase, into a form which is then implemented. If at each stage the system is described *informally* (perhaps in English with equations and diagrams) then these definitions are always open to misinterpretation. If, however, these definitions are written in a formal specification language, their meaning is totally unambiguous and precise. Formal methods can therefore be used to reduce ambiguity and to enhance our understanding of the system's operation. Also, if the system is defined formally at each stage, it should be possible to *prove* the equivalence of the different descriptions of the system and thereby verify that the transformations have been performed correctly. If this is done across all the stages of development it allows the designer to prove that the final design satisfies the original specification.

Unfortunately, the use of formal methods is not perhaps as simple as the last paragraph might suggest. While the use of formal specification languages is straightforward, the process of proving the equivalence of different stages of the design is a very specialised task requiring a high degree of mathematical ability. For this reason, very few projects use formal methods throughout the entire development lifecycle. However, formal methods are widely used within certain phases of the development process and engineers should be aware of these tools. For an introductory overview of formal methods readers are referred to the relevant chapter of [Storey 1996].

5.4 Managing the Design Process

Safety is not achieved simply by following an appropriate set of development methods. It must be planned and built into a product by considering safety at all stages. Because of its importance it is essential that a *safety culture* is encouraged *from the top* of the organisation. A well defined safety policy is needed to establish working practices and to ensure that these are followed.

While management structure will inevitably vary from one company to another some well defined hierarchy must be defined to allocate safety responsibilities. One of the key features of this structure should be the incorporation of an appropriate level of independence between the various roles. Independent verification of design tasks greatly increases our confidence in their correctness and therefore enhances safety. The various standards (such as IEC 1508) give guidance on the degree of independence that is appropriate for a range of tasks for systems of different levels of integrity.

References

[Anderson 1990] Anderson T, and Lee P A: *Fault Tolerance: Principles and Practice*, 2nd edn., Springer-Verlag, New York, 1990.

[Avizienis 1985] Avizienis A: The N-version approach to fault-tolerant software, *IEEE Trans. Software Eng.*, **11**(12), 1491-1501, 1985.

[Carré 1990] Carré B A, Jennings T J, Maclennan F J, Farrow P F and Garnsworth J R: *SPARK - The SPADE Ada Kernel,* 3rd edn, Program Validation Limited, Southampton, 1990.

[DoD 1992] *Military Standardization Handbook: Reliability Prediction of Electronic Equipment,* United States Department of Defense MIL-HDBK-217F, 1992.

[IEC 1995] Draft International Standard 1508 *Functional Safety: Safety-Related Systems*, International Electrotechnical Commission, Geneva, 1995.

[Jackson 1983] Jackson M: *System Design*, Prentice-Hall, Englewood Cliffs, NJ, 1983.

[Leveson 1995] Leveson N G: *Safeware: System Safety and Computers* Addison-Wesley, Reading, MA, 1995.

[Lewis 1996] Lewis E E: *Introduction to Reliability Engineering* 2nd edn., John Wiley, New York, 1996.

[Littlewood 1993] Littlewood B and Strigini L: Validation of ultrahigh dependability for software-based systems, *Comm. ACM,* **36**(11), 69-80, 1993.

[MISRA 1998] *Guidelines for the use of the C Language in Vehicle Based Software,* Motor Industry Software Reliability Association, Nuneaton, UK, 1998.

[Pyle 1991] Pyle I C: *Developing Safety Systems: A Guide Using Ada,* Prentice-Hall, Hemel Hempstead, UK, 1991.

[RSRE 1987] *The Official Handbook of MASCOT* Version 3. RSRE Computer Division, Malvern, 1987.

[STARTS 1987] STARTS Purchasers' Group: *The STARTS Guide: Vol 1,* 2nd edn., National Computing Centre Publications, Manchester, 1987.

[Storey 1996] Storey N: *Safety-Critical Computer Systems* Addison Wesley, Harlow, UK, 1996.

[Wichmann 1993] Wichmann B A: Microprocessor design faults, *Microprocessors and Microsystems,* **17**(7), 399-401, 1993

[Yourdon 1979] Yourdon E and Constantine L: *Structured Design: Fundamentals of a Discipline of Computer Program and Systems Design* Prentice-Hall, Englewood Cliffs, NJ, 1979.

Experiences with Safety Case Documentation According to the CENELEC Railway Safety Norms

Øystein Skogstad
SINTEF Telecom and Informatics
Trondheim, Norway
e-mail: Oystein.Skogstad@informatics.sintef.no

Abstract

In the assessment of safety aspects for the Gardermobanen railway signalling system, the methods indicated in the CENELEC Railway safety norms were utilised. The safety documentation was presented as Safety Cases that were assessed. During the development and engineering period, a proactive way of working was used by the Assessor, to disseminate understanding of the Assessor's point of view as early in the process as possible. Thus, the Assessor's sources of information were taken from the development process, the system descriptions as well as early drafts of the Safety Cases.

1. The project

In October 1998 Norway opened its new main airport, the Gardermoen Airport, located in the Oslo area. To establish a good public transport connection to the airport, it was decided to build a new railway line; The Gardermobanen line. The length of the line is about 40 km. The line speed is 210 km/h.

To control the train movements, it was decided to use a SIMIS system based signalling system, delivered from Siemens AG, Braunschweig. SIMIS is the name Siemens use for their basic (safe computer) technology. The signalling system consists of functional units for centralised traffic control, local operation, interlocking and automatic train control (ATC system). One sub-supplier is Ansaldo, who delivered the ATC transponders and their controllers. The transponders consist of electronic modules for detection of train movement and transfer of safety (signalling) information to the on board ATC equipment. For the transport of airport passengers a special airport train was purchased.

The interlocking system is a modification of a well-established system, proven in use. Many safety authorities in other countries have approved this system. The modifications decided upon were carried out on the application software, to adhere to specific Norwegian rules and regulations. It is common in Europe, that each country has its own peculiarities regarding how to safely operate railways. Thus, there is generally a need for nation specific software or data modifications to adhere to specific rules and regulations.

The trackside objects as well as the transponders are standard equipment. Each transponder has its individual data about speed limits etc., depending on where on the line it is situated. The ATC system is a new development, both hardware- and software.

The system at hand is a rather large system. The vital parts include approximately 50 processors (including duplications), some 600 signals (lights), 300 points, 400 track circuit and some 2000 transponders.

It was decided that SINTEF should be safety Assessor for assessing generic system safety aspects as well as installation specific aspects. A process with close co-operation between the customer side, the authority side and the Assessor was established.

2. Essential Safety Requirements in the CENELEC norms - and their Implementation

In the requirement specification for the signalling system, adherence to the CENELEC safety oriented railway norms was laid down. The norms in question are prEN 50126 [1], prEN 50128 [2] and ENV 50129 [3].

2.1 The norms

The three norms sets out requirements for all RAMS (Reliability, Availability, Maintainability, Safety) aspects:

- prEN 50126 relates to railway systems in general. It defines a systematic process for the specifications of RAMS requirements and the demonstration that these requirements are achieved.
- prEN 50128 relates to the software parts of the signalling systems. Requirements and recommendations for methods and techniques to be used throughout the software lifecycle are given.
- ENV 50129 relates to signalling systems. It sets out requirements for the contents of the Safety Case, and for the safety approval process. Furthermore it defines the interpretation and use of Safety Integrity

Levels (SIL). The standard in addition contains information for identifying the credible failure modes of hardware components, recommendations of techniques to be used for avoidance of systematic faults and control of random and systematic faults.

2.2 The implementation

The norms constitute some hundred pages of requirements, guidelines and other information relating to the safety of signalling systems. In the implementation of the norms, different detailed practices evolve in different organisations. The differences may have to do with the interpretation of the norms, as well as with how the norms are used. This reflects the way different organisations prefer to implement practical solutions to the sometimes general phrases of the standards.

Some issues that were emphasised in the assessment of safety for the Gardermobanen line signalling system were:

- Hazard Analysis
- Hazard Log
- Safety Calculation
- Requirements Traceability
- Process compliance

In the following paragraphs, we will give a summary of some of the discussions on these topics in the Gardermobanen project, as well as some general experiences from other projects.

2.2.1 Hazard Analysis

In principle the Hazard Analysis may be implemented as a Preliminary Hazard Analysis (PHA), a System Hazard Analysis (SHA) and an Operational Hazard Analysis (OHA). The PHA should be carried out during the concept phase. The hazard analysis and risk assessment processes defined in EN 50126 are necessary, in order to identify the required level of Safety Integrity for each particular situation. For interlocking as well as for the train control applications, the PHA was not carried out. It was instead decided to consider all system parts as high integrity parts (SIL 4). Such situations do tend to occur, when "new" standards (like the norms) are applied to "old" equipment types, and "old" applications.

The SHA and the OHA were carried out using a FMEA-type of analysis technique. The results from the SHA and the OHA were presented as "imposed conditions" (the proper denomination would have been "Safety-related application conditions"). These are the rules, conditions and constraints that shall be observed in the application of the system/sub-system/equipment. These conditions were presented by the supplier's safety manager, as an important part of the Safety Case documentation for the Gardermobanen signalling system.

2.2.2 *Hazard Log*

It is difficult to achieve support for using a Hazard Log in the system development or system modification phase. Usually, it is easier to get support for it in the operational phases (including commissioning). In the operational phases, the Hazard Log can be achieved through some field reporting scheme, where each failure in the field is categorised as safety relevant or not. In the reporting system, an equipment error can be reported as safety relevant. If so reported more thorough actions were taken, in addition to remedying the fault. Often such actions lead to reconstruction of some system part. It is very difficult to detect software errors this way. To detect software errors one should combine the traditional equipment error reporting system with some system recording operational anomalies.

2.2.3 *Safety Calculation*

System related safety calculations were carried out for the Gardermobanen signalling system. The purpose of such calculations is to demonstrate the possibility of fulfilment of the safety target values. Usually such calculations are carried out in a rather optimistic way: Software is considered error free, and few (if any) coverage factors are introduced in the calculations, to model non-ideal redundancy switchover. The calculations are usually system oriented, not application oriented. Therefore comparisons between different systems may be difficult, if the system structures are very different.

2.2.4 *Requirements traceability*

Requirements should in general be traced down to implementation. This mirrors the functional partition of the system. Furthermore, requirements should be traced to test-specifications. This is not always done, as some suppliers and administrations do not put enough effort into system (requirements) specification. In such cases, the well-known "battle about the system functions" does tend to occur under test and commissioning.

In the Gardermobanen case, the adaptation to specific Norwegian rules and regulations affected only a minor part of the system. Therefore, it was considered unnecessary to establish documentation of requirements traceability for the total system. A trace was done as a mapping between the requirements specification and the test specifications. The trace was not very detailed, however, and established rather late in the project. Therefore some discussion about functionality did arise during commissioning testing.

2.2.5 *Process compliance*

Every system development and delivery is accompanied by huge volumes of documents defining plans for the work, like a development plan, a quality plan, verification and validation plans, test plans, configuration management plans and even more. Some times the plans may be in accordance with the principles laid down in the norms, or at least in accordance with what may be judged as sound engineering practices. This does not necessarily mean that the plans are followed in practice. Adherence to approved plans should be an important topic when assessing whether approved principles for development and delivery of a safety related product has really been achieved. To assess this, the assessor must visit the supplying company, and do some investigations of how the work at the company is really carried out. One way to do such an investigation, is through some safety-oriented audits. In case the plans are not followed, or in case the implementation of the plan drifts too far away from the approved principles, safety related non-compliance statements may be issued by the assessor. This will in most cases lead to some actions by the company, to remedy the situation.

3. Software aspects

3.1 Software structure

The SIMIS system is realised with a simplex software solution. Duplicated hardware processors are running identical software programs. A commonly used solution is a 2 out of 2 duplication, with another pair in hot standby. A software failure may thus affect the total functionality of a processor pair, and with some probability the spare pair. Such software failures are usually state dependent. The remedying solution to such failures is restart of the processors. This solution is often deemed satisfactory, in situations dominated by normal hardware component failures. The defences utilised against software failures in the SIMIS system are:

- Defensive programming, like error traps, careful use of jumps, no loops and the like.
- A system that is proven in use. When a software system is well proven, and the user profile for the software does not change, the proven in use argument is considered relevant.
- The new system installations are thoroughly tested. In our case, there are laboratory tests, as well as extensive site testing, before putting the system into operation.
- All software is checked in detail by an independent third party. For safety systems with non-diverse software, such checking is considered necessary. The methods used for such checking include structured code walkthroughs. In the walkthrough, the code is checked against the specification, and the functionality tested at the software unit level. The checks are performed by personnel with detailed knowledge of the application, and many years of experience with this particular type of application.

The software structure is a unit oriented structure. Some units may be global, like a library. Other units are local. Data is kept as local as possible, but global data is allowed.

The SIMIS system is also a distributed system. Processors dealing with geographical information are distributed. Even so, such computers are equipped with identical software, as long as they are doing the same types of tasks. The data (parameters) for the geographical computers differ from computer to computer, to cater for differences in the computer environment.

3.2 Development process

The basic SIMIS interlocking system is proven in use over several years. The modifications for the Norwegian implementation were necessary only for the functional modification. A team of experienced programmers, well acquainted with signalling system applications did the programming. Programming was performed in assembly code. According to the norms, assembly code is not recommended. However, there were some very general reasons for allowing assembly programming even for the new functions for Norway:

- not only the system is "proven-in-use", the same argument can be applied to the development methods that are used also for system modifications.
- the assembly code is highly structured, and well commented.
- the verification of the code is rather thorough, utilising application experience, the assembly code itself as well as some high level representation of the code.

Due to the modular structure, each unit can be tested independently for its functionality. The programmers performed these tests, before releasing the unit. Before delivery, the system was tested in the laboratory, with simulated environment. This test was carried out in two stages; first as an internal system test, thereafter as a Factory Acceptance Test.

After installation on site, a Site Acceptance Test was performed. The Site Acceptance Test was carried out with experienced and certified personnel from the customer. These persons were especially skilled in signalling system operation in general. They had received a thorough classroom and laboratory education on the system. These persons also took part in the Factory Acceptance Tests.

The system was delivered and put into operation in phases. Each phase represented a section of the line. In each new phase, new functionality was typically included. In addition, changes were often introduced in the next phase due to cases of minor non-compliances. The changes in functionality in a new phase were retrofitted on system parts already put into operation.

Thus, a disciplined attitude to regression testing was necessary. As a starting point, regression testing was not considered necessary, because the changes in software were to be deemed free from side effects. However, when anomalies were detected, the SAT testers had the liberty to do regression testing, on their own discretion. Some tests performed under FAT were not reproduced under SAT, due to practical reasons. It was accepted that software logic oriented tests would most certainly give the same results in simulated as in real environment. The SAT included classical railway signalling tests, like manipulations of objects, setting of train-routes etc.

An Error and Requirements list was maintained based on test results from FAT and SAT. The list got its name because not all criticisms from the testers were really errors. Changes in relation to the requirements were included as well. The list altogether consisted of some 700 items.

3.3 Independent validation

A special branch of the Supplier Company did code inspection and unit testing to check that programs and changes were free from unwanted side effects. Furthermore, they did check application specific safety aspects. This branch was certified by a national (German) safety body to act as an independent organisation. Thus, independent validation (as well as verification) according to the CENELEC norms was performed. The review reports from this organisation were deemed by the safety Assessor (SINTEF) as important evidence of technical safety. Therefore, these reports formed a major part of the Generic Safety Case.

3.4 Process related problems

To check compliance to EN50128, which is rather detailed on software development, it is necessary to gain a close understanding of the development processes of the supplier. Furthermore, it is necessary to understand the detailed product structure, as well as the relation between software product structure and hardware product structure. It is also important to understand the relation between the product structures and the development and engineering processes, with associated delivery milestones and validation activities. In a large project, these relations are rather complex. To understand them requires availability of descriptive documents from the supplier. Not all suppliers are necessarily keen to provide such detailed information to people outside the company. In the Gardermobanen project, mutual confidence was established through several meetings and discussions.

EN50128 has a concept for the software development process, and associated documentation. Newly established departments (or companies) tend to organise their documentation according to these concepts. Old and well-established companies do, however, have a history of how they organise things, that is not necessarily easily mapped to the EN50128 process concept.

4. The Safety Case

4.1 Requirements

The Safety Case shall contain all necessary information to assess safety of a system, an application or a specific installation. In addition, a good Safety Case will give information that makes the Assessor comfortable with the Reliability, Availability and Maintainability properties of the system. Safety should not be viewed separately from these attributes.

4.2 Contents

The CENELEC norms standardise the contents of the Safety Case. This makes it easier for a safety Assessor to find his way through Safety Cases from different suppliers.
The contents of a Safety Case shall be:

1. DoS - Definition of System
2. QMR - Quality Management Report
3. SMR - Safety Management Report
4. TSR - Technical Safety Report
5. RSC - Related Safety Cases
6. Conclusion

4.3 Types of Safety Cases

Three different types of Safety Cases are identified in the norms:

- **Generic Product Safety Case**. Generic products can be reused for quite different applications. A computer may be considered a generic product.
- **Generic Application Safety Case**, which is valid for a class of applications. Signalling systems in general may be considered as a class of applications.
- **Specific Application Safety Case**. A specific installation (instance) may be considered as a specific application.

The Specific Application Safety Case shall be divided into two portions:

- **Application Design Safety Case,** which is an evidence of safety for the theoretical design of the specific application.
- **Physical Implementation Safety Case,** which is a safety evidence for the implementation of the specific application.

For a new system, these rules are most often applied almost directly, as shown in Table 1. It should be noted that in practice General Application or Application Design Safety Cases might be established per country, due to the national Rules and Regulations.

In the Gardermobanen case, the system was a modification of an already proven-in-use system. This system being approved by several railway administrations and safety authorities. The development of the basic SIMIS system had started long before the CENELEC norms came into use. Thus, no Safety Cases had been produced for the basic system. As a compromise to the ideal situation presented in the CENELEC norms, we decided to adapt the picture to the situation shown in Table 2.

Type/instance	Type of Safety Case	Use of Safety Case
General system	Generic Product Safety Case.	Subject for cross acceptance.
General Interlocking system	Generic Application Safety Case.	Subject for cross acceptance.
National (type) Interlocking system	Application Design Safety Case.	Subject for Type Approval.
Specific (instance) interlocking system	Physical Implementation Safety Case.	Subject for putting into operation.

Table 1. System type/instance, Safety Case types and Use of the Safety Case. New system.

Type/instance	Type of Safety Case	Use of Safety Case
Proven-in-use interlocking system	No Safety Case. Documentation of approval by some notified body.	Cross acceptance of basic system.
National type interlocking system	Generic Safety Case (Product and Application (Design)).	Subject for Type Approval.
Track section (partial build). Complete line (total build).	Specific Application Safety Case (Physical Implementation).	Subject for putting into operation.

Table 2. System type/instance, Safety Case types and Use of the Safety Case. Old system.

4.4 Application

In the original planning of the safety approval process, the concept of type-approval was significant. Thus, initially the focus was on the Generic Safety Case that was planned used for type approval. However, already from the initial planning, a type approval was dependent on some field experience with the actual system. Thus, the Specific Application Safety Case was also needed for the type approval process. On the other hand, to have field experience, you need a field operation. And to get a field operation you need a safety approval. Normally, making the first installations on a trial line solves this catch. In our case, the first application would be the national prestige project of the commuter train for the new airport.

To solve the problem, both the Generic and the Specific Safety Cases had to come in several versions. The following main versions of Safety Case documentation were used:

- Pre-FAT version. This version was used to approve start-up of FAT activities.
- Pre-SAT version. This version was an update of the pre-FAT version, with the results from the FAT, including system changes. It was used to approve SAT start-up.
- As-built version (post SAT). This version of the Safety Case was the Pre-SAT version, updated with (references to) the test reports from the SAT. In practice, when there were system or configuration changes during SAT, the difference between Pre-SAT and post-SAT Safety Cases would include new versions of descriptions of functionality and of data. In the cases where the SAT was aborted before system change, a new pre-SAT version had to be made. In some cases, when "everything went wrong", the supplier had to make a new Pre-FAT version as well.

4.5 Detailed requirements on Safety Case documentation

The most important lessons learned through assessment projects are about the contents of the Safety Case. A specific project will always have special circumstances that have to be considered. This section is therefore not always a direct reproduction of the contents of the Safety Cases in the Gardermobanen project, but rather a synthesis of what in general may be the contents of a Safety Case.

The first consideration is that there are major differences between the Generic Safety Case and the Specific Safety Case, with regards to the contents. We will present opinions about the contents of both types, based upon the assumption that both types of Safety Cases are assessed according to the norms.

4.5.1 Definition of system

The definition of system shall precisely define or reference the system/subsystem/equipment to which the Safety Case refers, including version numbers and modification status of all requirements, design and application documentation. The contents of this part of the Safety Case should be as described in Table 3. The Table has the following structure:

- Information type, i.e. the type of information discussed.
- Opinion on whether the information type should be considered necessary in the Generic Safety Case.
- Opinion on whether the information type should be considered necessary in the Specific Safety Case.
- Comments to the Yes/No entries in the table.

Information type	Generic Safety Case	Specific Safety Case	Comments
System Description	Yes	No	
Interface overview (general)	Yes	No	
Product hardware Structure	Yes	No	Gives a systematic reference to all hardware (PCB) and equipment units of the entire product.
Product Software Structure	Yes	No	Gives a systematic reference to all software units of the entire product.
Baseline • general • installation	 Yes No	 No Yes	See below.

Table 3. Information in Safety Case /Definition of System

Baseline documentation

In our opinion, good baseline documentation shall be very specific. In the Gardermobanen case, the following information was produced:

- FAT/SAT relation (phase, site, ...).
- Description of changes since last baseline. Functional changes in the Generic Safety Case. Configuration and software data changes in the Specific Safety Case.
- Requirements (reference).

- Design documents (reference).
- Test protocols, test logs, test reports (reference).
- Hardware units included in the configuration (in relation to product structure).
- Software units included in the configuration, with version number, per computer (in relation to product structure).
- Chip version numbers and checksums (for all computer chips).
- Drawing number and revisions for all equipment and site drawings, including:
    - Track layout
    - Object placements
    - Transponder code tables.

  In practice, this may be done as a reference to a register of drawings.

4.5.2 *Quality Management Report*

The QMR shall contain evidence that a sound quality assurance process has been performed. Nowadays, it is virtually impossible to imagine such a process that does not comply with the structure and requirements of the ISO 9000 series of Quality Management and Quality Assurance standards. The topics covered in these standards are usually described in a quality assurance plan. There should be quality assurance plans for

- the product development or modification.
- the engineering and installation/testing work to be performed in a delivery.

In the Gardermobanen project, the following information was paid attention to for the Generic Safety Case:

- Company quality management policies and plans.
- Quality Assurance Certificates (ISO9001 or ISO 9002 certificates).
- Register/description of QA models and methods.
- Quality audit non-conformances and their current status (for the modification work as well as the production and installation work).
- Register of reports from review or inspection work performed. The purpose of this information was to ensure that all planned, imposed or implicit reviews and inspection had been carried out. The types of reports will normally be different for development, production and installation work performed.
- Register of phase related supplier management reviews.
- Type certificates for modules and units (hardware), according to hardware product structure. These certificates should contain information about environmental robustness, including EMC, temperature, shock and vibration.
- Supplier's conclusion.

For the Specific Application Safety Case, the following information was emphasised:

- Register/description of specific QA models and specific methods (i.e. models and methods not described in the Generic Safety Case)
- Installation test report.
- Specific site review reports. This should include the engineering and data reviews.
- FAT/SAT test reports. The test should include configuration checking.
- Supplier's conclusion

For some reason, it seems to be difficult for the supplier to write a good conclusion, both for the QMS and for the SMR (see below). The QMR (and SMR) report tends to include only a synopsis of the activities carried out. A conclusion in form of an analysis of why the performed activities are sufficient, is harder to find.

4.5.3 Safety Management Report

The safety management work shall be done according to the safety plan. This plan should be established, and agreed upon before the Safety Cases are produced. For reference purposes one might anyhow include the Safety Plan in the SMR.

The CENELEC norms say a lot about what shall be the constituents of the safety plan, like safety organisation, safety lifecycle and planned safety activities. The norms are quite extensive in this area. The contents of the SMR should give evidence that the planned activities as defined in the safety plan, are all carried out. In the Gardermobanen project, special attention should be given the topics as shown in Table 4.

4.5.4 Technical Safety Report

The chapter structure of the technical safety report is normative. In practice, the norms only describe the contents of the Generic Safety Case TSR.

The norms prescribe the following main chapters of the TSR:

4.1 Introduction
4.2 Assurance of correct functional operation
4.3 Effects of faults
4.4 Operation with external influences
4.5 Safety-related application conditions
4.6 Safety Qualification Tests

This structure is well suited for the Generic Safety Case. It is stated in the norms that "The Technical Safety Report shall explain the technical principles which assure the safety of the design, including (or giving references to) all supporting evidence (for example, design principles and calculations, test specifications and results, and safety analyses). Large volumes of detailed evidence and supporting documentation need not be included, provided precise references are given to such documents."

Information type	Generic Safety Case	Specific Safety Case	Comments
Register of competence	Yes	No	Includes any accreditation of personnel or organisation doing independent validation (third party validation).
Register/description of safety models and specific methods	Yes	No/Yes	In the specific SC is only mentioned additional methods not described in the Generic SC.
Register of safety related internal and external audits, status for issues in audits	Yes	No	
Register of safety reviews/walkthrough on changed drawings, software,	Yes	Yes	In the specific SC is only mentioned additional reviews not referred to in the Generic SC.
Hazard Analysis PHA SHA OHA	 Yes Yes Yes	 No No No	 In all the analyses Imposed Conditions should be explicitly presented.
Hazard Log	Yes	No	
Safety/Reliability calculations	Yes	No	
Explicit imposed conditions from Hazard Analysis / Hazard Log	Yes	Yes	
Supplier's conclusion	Yes	Yes	

Table 4. Contents of the SMR

Some very detailed checklists (mandatory lists) for the contents of the sections above are given in the norms. A supplementary view of what might also be the contents of the TSR is given in Table 5, based upon the experiences from the Gardermobanen project. The table includes some opinions on the content of the Specific Safety Case TSR.

Information type	Generic Safety Case	Specific Safety Case	Comments
Description of technical characteristics of the system. According to structure from CENELEC norms.	Yes	No	According to the normative chapter structure, as described above.
Validation reports on requirements.	Yes	Yes	System and general process requirements in the Generic Sc. Specific (installation related) requirements in the Specific Safety Case.
Validation reports for hardware modules.	Yes	No	If applicable. According to hardware product structure.
Validation reports for software units.	Yes	No	According to software product structure.
Safety certificates for modules and units (hw, sw).	Yes	No	Safety Certificates for modules may be • a validation report, • a specific Safety Case, • Module approval by some authority.
Validation reports for all safety relevant tools.	Yes	No	
Validation reports for all site specific documentation, track layouts, object placement and coding and other projecting data.	No	Yes	With projecting data is meant all information about the track engineering work.
FAT/SAT rest reports.	No	Yes	
Supplier's conclusion	Yes	Yes	

Table 5. The contents of the Technical Safety Report.

In the table is the term Validation Report used. A validation report in this context should emphasise safety. It should include:

- A report from safety related review, walkthrough and the like.
- An analysis of safety aspect of the subsystem/module at hand. Potential hazards should be included in the Hazard Log.
- Specific safety related testing of unit at hand. For a software unit, this should include testing of unit behaviour with input outside specified values.

4.5.5 *Related Safety Cases*

The section on Related Safety Cases should give references to Safety Cases related to subsystems, if applicable. In some situations, system modules might have certificates related to safety associated with them. This may relate to hardware modules, e.g. off the shelf modules. There should be references to Safety Cases for other vital systems that constitute a functional totality of some kind. In the Gardermobanen Case, a natural division was separate Safety Cases for the CTC, system, the interlocking system, the ATC system and the transponders. In such cases a separate Safety Case may be needed, to show safety assurance of the system interworking aspects.

4.5.6 *Conclusion*

It seems to be difficult for the supplier to write a good conclusion on the Safety Case. The conclusion tends to include only a synopsis of the activities carried out, and a short description of system attributes that contribute to safety. A conclusion in form of an analysis of why the performed activities and the system attributes are sufficient is harder to find.

5. The Safety Assessment Process

5.1 Introduction

SINTEF was appointed safety Assessor, both for the type-approval aspects and for the actual implementation of the signalling system at the Gardermobanen line. In the work, SINTEF used three viewpoints, to cover all aspects of safety. These viewpoints were:

1. System oriented.
2. Process oriented.
3. Safety Case oriented.

5.2 System orientation

When presented with a new system for assessment, it is usual for an Assessor to go through the technical description of the system in some detail. The purpose of this is:

- To understand the functional capabilities of the system.
- To understand the system design. The following design aspects should be checked:
  - Hardware structure, including redundancy.
  - System build-up in racks and cabinets, including earthing principles, shielding, EMC etc.
  - Software structure.
  - Communication protocols, at high and low level (Communication and interface structure). This includes the understanding of applied CRCs or any similar type of communication codes.
  - Reliability models and calculations.
  - FTA, FMEA and other safety or reliability analyses of the system.
- To raise more detailed questions, if any part of the documentation is out of date, obscure, contradicting or lacking.
- To understand the failsafe properties of the system. This may be implemented in form of redundancy switchover, defensive design, utilisation of error detection and error correction techniques (including codes), etc.
- To understand how the system at hand implements the national Railways Rules and Regulations.

In addition to going through system descriptions, systems- and tools demonstrations and other forms of communication with the system development team is considered useful, and often necessary.

5.3 Process orientation

SINTEF performs walkthroughs of all process-related documentation, like quality assurance plans, safety plans, V&V plans and other plans related to system safety. Comments to the processes applied by the supplier are given. To understand how the work is carried out, we use safety-oriented audits as well as visits and meetings with both supplier and customer and other involved parties. In the audits, we issue safety-related non-compliance reports, when necessary.

The process parts most often overlooked by both customer and supplier, are the early project phases (concept, etc). We do not always require documentation of the work from these phases. Some times this documentation does not exist. The documentation tends to start with the requirement specification. One reason for this may be that the parties involved have long experience with signalling systems, so the topics that should have been addressed in the earlier phases "go without saying". Depending on the situation, this could be deemed as satisfactory, from a safety viewpoint. Special emphasis is directed towards understanding the methods, techniques and tools for software development.

When the system is put under test, we participate as an observer in some of the tests. This may be especially interesting tests, or we may participate on a sampling basis. SINTEF looks into the safety assurance aspects of the engineering (projecting) works. With regard to the definition of train-routes, signal placement etc., we do some checks to verify that sound engineering practices have been used.

In our experience open and continuing dialogues between the parties involved give the Assessor the possibility to state his viewpoint early in the process. We often use the term "Proactive" to illustrate this way of working. However, care should be taken that the Assessor does not prescribe particular solutions.

5.4 Safety Case Orientation

The Safety Case is suitable for accumulation of both product and process documentation. In some cases a long time period is needed, for supplier, customer and Assessor to agree upon a satisfactory content of the Safety Case. Therefore, not to waste time, the draft Safety Case should be presented early in the project. This will give all parties the possibility to understand the contents of the Safety Case, and to comment early on substandard sections in the Safety Case.

Due to the product and process investigations carried out, one may have to accept a system put into service without a perfect, or complete, Safety Case. The Safety Case should be considered *documentation* of safety aspects already taken care of and known by the assessor. If the gap is too large between what the Safety Case is and what it should have been, system putting into operation may have to wait until the documentation is completed. With too much information lacking, there is a risk that some safety relevant aspect is overlooked.

The Safety Case in either of its phases (pre-FAT, pre-SAT, post SAT), is evaluated by the Assessor. Comments are given, and the Assessor gives a recommendation with regard to putting the system into operation (if applicable).

A special problem in the Gardermobanen project, was the stepwise putting into operation of sections of the tracks. The first geographical area was put into operation early in 1997. The last area was put into operation in the second half of 1998, except for the tunnel part that is postponed until 1999. In this period, new sections have been put into service in steps. In addition, the functionality of the system has been altered, to cater for the items on the Error & Requirement list.

Each system change needed to be mirrored in the Safety Case for the system. Prior to putting a new track section into operation, the Safety Cases needed to be updated to mirror the current situation. This has lead to a substantial number of revisions (more than 10) of the Safety Cases. Each revision needed to be checked if it was good enough to defend the putting into operation with the new functionality or the new section. This rather energy consuming way of working was decided upon, in an effort to save calendar time from project start-up until final opening of the airport train service.

Separate Safety Cases were submitted for the main system parts; Interlocking, ATC and CTC. This gave in three Generic Safety Cases, as well as three Specific Safety Cases. But in practice there was a tendency towards three Generic Safety Cases plus three Specific Safety Cases *pr. track section,* until everybody learned to incorporate information from all sections even if only one new section was opened.

6. The Safety Assessment Report

For the Gardermobanen project, it was not practical to produce a Safety Assessment Report for each section of the line - especially since each track section oriented situation existed for a very limited time. Therefore, for each track section (or other major system change), the Assessor produced some detailed comments to the Safety Cases, concluding with recommendations on:

- Whether the Safety Case could be considered as a complete and correct documentation of system safety
- Whether possible inaccuracies or omissions in the Safety Case were of such a character that putting the change or section into operation should be advised against.

This way of working was made possible through:

- Regular information meetings between the parties involved.
- The supplier's timely submission of all relevant information to the parties.
- The Assessor's possibility to immediately follow up on whatever item that seemed unclear or otherwise safety relevant.

Thus, the Assessor continually had to have a good understanding of what was going on in the project.

The final safety assessment report will report the conclusions, when the intensity of system changes has dropped to a lover level, and the system is in stable operation and maintenance. The norms include requirements for the Safety Assessment Report. The report should explain the activities carried out by the Safety Assessor to determine how the system has been designed to meet its requirements, and possibly specify some additional conditions for the operation of the system/sub-system/item of equipment.

A possible TOC for the Safety Assessment Report for the Generic System is:

1. Summary and conclusion
2. Definition of system
3. Assessment of quality management
    - 3.1 general quality system
    - 3.2 system development plan
    - 3.3 software quality assurance
    - 3.4 reviews and inspections
    - 3.5 problem reporting
    - 3.6 change control, and configuration management
    - 3.7 quality records
    - 3.8 assessment
4. Assessment of safety management
    - 4.1 safety life cycle
    - 4.2 safety organisation
    - 4.3 safety plan
    - 4.4 hazard analysis
    - 4.5 hazard log
    - 4.6 safety requirements specification
    - 4.7 requirements traceability
    - 4.8 safety reviews
    - 4.9 safety case
    - 4.10 safety verification and validation
        - 4.10.1 safety analysis and calculation.
    - 4.11 system/subsystem/equipment handover
    - 4.12 operation and maintenance
    - 4.13 decommissioning and disposal
    - 4.14 safety calculations
    - 4.15 audits conducted by sintef.
    - 4.16 assessment
5. Assessment of technical safety
    - 5.1 introduction
    - 5.2 assurance of correct functional operation
    - 5.3 effects of faults
    - 5.4 operation with external influences
    - 5.5 safety-related application conditions
    - 5.6 safety qualification tests
6. Related safety cases
7. Conclusion

The same overall structure is kept also for the Safety Assessment Report for the Specific System, but the details are of course different.

7. Conclusion

The main points that may sum up the utilisation of the CENELEC norms as a base for safety assessment, and especially the use of the Safety Case as the "point of investigation", are presented below.

- It is never enough to do only a paper study, comparing the norms with some papers from the supplier stating what the supplier (might) have done on the various requirements of the norms. The assessor needs a more detailed level of understanding, together with assurance of how the implementation of the requirements is really carried through.
- Some of the requirements or recommendations of the norms are rather general. Some may even be a little out of line with modern development practices. It is therefore essential that the supplier and assessor learn to see eye to eye on what are the sound interpretations of general requirements and recommendations.
- Time is needed to establish by the supplier a form of the Safety Case, that is concrete and detailed enough to be of any use. The supplier does have a strong tendency to write Safety Cases with as little detail as possible.

Once a good Safety Case with enough detail is established, it serves as a very good means for communicating safety aspects relating to the system, or some specific installation.

The application of the norms on old systems should be considered carefully. The development methods and techniques will most certainly not adhere to the recommendations of the norms. But the system is proven in use, so "by some magic" it has been possible to achieve safety without adhering to the norms. Above all, the safety documentation is usually not as explicitly present as presumed in the norms. The supplier's build-up of a Safety Case helps on the last problem in a wonderful way. The situation is very often that the particular safety aspect is well taken care of, for the elder system, but under quite different names and without explicit documentation directed towards a third party.

Even for elder systems the suppliers could turn to the norms, and document their compliance with the norms. The advantages of this are numerous. The supplier gets the possibility for a detailed and structured walkthrough of his safety activities. Organisation and responsibilities will become much clearer, even to the parties involved. Documentation will improve, and be commensurable with documentation from other suppliers, thus easing the introduction of the product to new marked groups.

Acknowledgement

I would like to acknowledge the valuable contributions from the persons taking actively part in the development- and acquisition processes as well as commenting on the drafts of this paper. A special acknowledgement to the project Safety Manager at Siemens; Hr. Wolfgang Bauch-Gelleszun for effective development of the Safety Case documents, and for our discussions with regard to the content of the Safety Cases.

8. References

[1] prEN 50126. Railway applications - The specification and demonstration of Reliability, Availability, Maintainability and Safety (RAMS). CENELEC

[2] prEN 50128. Railway Applications: Software for Railway Control and Protection Systems. CENELEC

[3] ENV 50129. Railway Applications: Safety related electronic systems for signalling. CENELEC

Retrospective Collection of Safety Case Evidence - Experiences from an Air Traffic Control System Safety Case

Tim Boyce, Vivien Hamilton

GEC-Marconi Research Centre

Great Baddow, Chelmsford, UK.

Abstract

This paper describes some aspects of the collection of evidence to support a retrospective software safety case for a large, military Air Traffic Control system. The evidence was required to justify the derivation of software specific safety targets and to demonstrate that the targets were met. Operational records of incidents, including 'near misses', and configuration and change management records provided the data used for the evidence. The paper concludes that where equipment has been used within an existing safety culture it is potentially possible to find enough evidence for a successful retrospective safety case.

1 The Challenge

1.1 Background

The challenge was to produce a safety case for the software at the heart of an existing Integrated Command and Control System (ICCS) which is a major part of the Royal Air Force's (RAF) UK Air Defence Ground Environment (UKADGE). The prime mission of UKADGE is to support the RAF in its provision of an air defence capability for the UK during times of tension and war. During peace time UKADGE supports the RAF in its training and preparation for such wartime requirements.

UKADGE comprises a complex network of control centres, radars (military and civilian) and radios. It has links to other European ground based air defence agencies, and airborne and shipborne control points. Along with the equipment the ground environment also has a significant human element, consisting of controllers, fighter marshals and systems managers, to name but a few.

In peace time, RAF operations require military aircraft to share parts of the UK airspace with civilian aircraft. Some of this airspace is classed as 'controlled airspace', requiring its users to conform with certain Air Traffic Control (ATC) requirements. This occurs, for example, when aircraft operate in the controlled airspace above 24,500 feet or transit the airways system that support traffic between the continent and the UK. It is these ATC requirements on air defence operations that mean that UKADGE, and consequently ICCS, has a safety related role.

The ICCS software has a long history of evolution over the twelve years since its first incarnation. The software in its current form has evolved from a baseline set five years ago. Over this period the software has undergone many changes due to modification and fault correction. The software comprises an estimated three million lines of code, written in various languages, which is distributed over three main sub-systems within each control centre.

Although safety considerations were an integral part of the process of software modification prior to the MOD requirement for a safety assessment, system and software safety had not been systematically addressed. This meant that there was no existing basis against which the safety of the system or software could be judged.

The initial requirement to provide a safety assessment (and subsequently to produce a safety case) of the software, therefore, proved to be a much greater challenge than initially envisaged. This was because of the need to carry out a much broader analysis, covering the whole context of UKADGE flying, to determine suitable high level targets against which the software could eventually be assessed.

1.2 Types of Evidence Required

The first part of the safety assessment of the software was to determine the safety targets against which the software could be assessed: the safety team had not only to gather evidence of compliance with targets, but also evidence to justify the targets used.

In general, retrospective safety assurance is difficult because of a lack of opportunity to influence the outcome, for example by adding safety features. However, retrospective safety assessment does have the potential advantage that records of product performance over long periods of use may exist from which evidence of adequate safety performance can be justified. Unfortunately, the fact that the data will not have been collected specifically for safety analysis means that effort is required to extract relevant and meaningful information.

A good approach to most forms of analysis is to consider what goal the analysis is trying to achieve, to derive a set of questions which need to be answered to show that this goal has been achieved, and finally to determine the data (metrics) which need to be collected to provide the answers to the questions.

In the retrospective analysis performed for the ICCS software safety assessment, the starting point was the available data (metrics) and the desired goal. It was then necessary to determine the questions that could be answered by the data that would demonstrate achievement of the desired goal.

2 Safety Targets

The first problem to overcome during the ICCS software safety assessment was a lack of relevant safety targets for ICCS software. Safety targets did not even exist for the UKADGE system as a whole.

The first of the following two sub-sections describe the aim and approach taken to deriving relevant safety targets for the ICCS software. The second sub-section describes the evidence used to justify the derivation.

2.1 Safety Target Setting Approach

The first problem was to derive safety targets directly applicable to the ICCS software.

The starting point for this work was the need to arrive at safety targets from established policy, thereby ensuring their widest acceptance. Although there is no single document to reference, the UK Civil Aviation Authority's (CAA) policy on the tolerability of accidental loss of commercial aircraft is implicit in their requirement that new aircraft are certified against a target of 1×10^{-7} losses per flight hour due to any cause. Military policy on accidental aircraft loss was taken from JSP 318B [JSP318B 1996] which defines safety targets for transport and other (non-transport) aircraft of 1×10^{-7} and 1×10^{-6} losses per flight hour respectively. However, the JSP 318B targets are specifically for accidents due to technical causes (i.e. airworthiness related).

The safety targets defined by these policies were not specific enough to be directly applicable in the safety assessment because they cover accidents to which UKADGE and ICCS cannot contribute. Hence a proportion of the policy targets for the particular accidents relevant to ICCS had to be derived.

Hazard identification and analysis established that the main accident type related to ICCS is mid-air collision. The main purpose of ATC is to ensure that aircraft remain at a safe distance from each other. Hence loss of, or incorrect, ATC are high level hazards which could result in aircraft becoming too close and, in the worst case, colliding. The ICCS software is responsible for providing both the radar picture by which the air traffic situation is monitored and the ground-to-air voice communications through which ATC instructions and information are passed from controller to pilot. Hence, ICCS software failure could result in the high level ATC hazards described earlier. The hazard analysis also highlighted that, for many potential equipment failures, the human operators provide substantial mitigation against the high level hazards.

It was also necessary to consider the different groups of people who could potentially be involved in mid-air collisions and to derive targets for each of the possible mixes of aircraft groups in such accidents (e.g. UKADGE military with Commercial Air Transport (CAT) or UKADGE military with General Aviation).

The safety principle of As Low As Reasonably Practicable (ALARP) was applied to the ICCS software safety assessment. The ALARP principle is described in [HSE 1992] and is embodied in a number of safety standards, including DEF-STAN 00-56/2 [MOD 1996]. In practical terms the application of the ALARP principle requires that two safety targets are defined; an upper safety target representing the absolute minimum level of safety that is tolerable; and a lower safety target representing a level of safety which is broadly tolerable. Between the upper and lower bounds is the 'ALARP region', within which the level of safety achieved is only tolerable if the cost of reducing the risks further outweigh the benefit gained. Hence, the safety target setting process involved defining and justifying an ALARP region.

It was decided that for the purposes of monitoring the performance of the ICCS software against the safety targets it would be better to define the targets in terms of relevant near miss events. Near misses occur more frequently than accidents and are thus a more practical indicator of safety performance. The near miss events relevant to mid-air collision accidents are minimum separation infringement incidents (i.e. when aircraft fly closer to each other than the prescribed criteria for safe separation). These incidents are referred to as 'minseps' for the remainder of this paper.

The final stage in deriving quantitative safety targets for the ICCS software was to determine the minsep target to apply to UKADGE as a whole and then to determine a suitable proportion of the UKADGE minsep target to apply to the ICCS software.

In addition to the quantitative safety targets, the risk classification scheme from DEF-STAN 00-56/2 was used to derive targets and to define the safety integrity level (SIL) requirements for the software.

2.2 Safety Target Setting Evidence

The following parts of this sub-section describe the evidence collected for the three main stages of the target setting process; apportionment of overall policy safety targets to UKADGE accident specific targets (encompassing the ALARP principle); the conversion of accident targets in terms of mid-air collisions into targets in terms of minseps; and the apportionment of the minsep targets to the ICCS software.

Although safety targets were derived for all mixes of aircraft groups which could potentially be involved in a mid-air collision with a UKADGE military aircraft, the remainder of this paper will only discuss the evidence for collisions involving UKADGE military with CAT and non-UKADGE military aircraft.

2.2.1 Mid-Air Collision Accident Targets and ALARP

The initial problem in defining specific, mid-air collision targets for military aircraft was that policy targets are only defined for losses due to aircraft technical causes. It was first necessary to derive a higher level target for loss due to any cause. For this purpose, data was needed to determine the proportion of all military accidents which are due to technical causes and hence justify an overall safety target consistent with the stated policy.

The data used in this justification was taken from the Inspectorate of Flight Safety (IoFS) database. This is an electronic database containing records of all accidents that result in damage to an aircraft. The database was queried for the number of accidents in which the aircraft was lost and aircraft technical problems recorded as the cause. It was estimated that technical causes should represent only small proportion (i.e. less than 10%) of the overall loss rate of military aircraft.

Data from the IoFS database was also used to justify the setting of a proportion of the overall safety target to be applied to mid-air collisions. However, there was a need to distinguish between two types of mid-air collisions; 'stranger' collisions and 'non-stranger' collisions. A 'stranger' collision is one in which the aircraft involved would normally expect some level of safe separation criteria to be applicable, and where ATC might be expected to play a part. A 'non-stranger' collision is one in which the aircraft involved are aware of each other's close proximity (e.g. during formation flying or air combat training where the risk of mid-air collision is higher). For 'non-stranger' accidents, standard ATC safe separation criteria do not apply and hence these incidents are not relevant to ICCS.

The database could only be searched for mid-air collisions and it was not possible to distinguish between these two types of mid-air collision automatically during the search. Hence it was necessary to manually sort the set of records into the appropriate collision categories.

The difficulty of determining a reasonable proportion of the CAT policy safety target for mid-air collisions involving military aircraft is because there have been no mid-air collisions between military and CAT aircraft in the UK. In lieu of actual experience, the expected rate of mid-air collisions was estimated by examining UK AIRPROX records and counting the number of minsep incidents involving these two groups of aircraft.

Combining this with an estimate of the likelihood of a collision occurring after a minsep incident (discussed in section 2.2.3) established that CAT - military mid-air collisions should be responsible for only a very small percentage of all CAT accidents.

2.2.2 *ALARP*

In determining ALARP targets it was assumed that the policy safety targets represented the lower bounds of their respective ALARP regions (i.e. the boundary between ALARP and broadly tolerable). This was considered to be a reasonable assumption since the policy targets are defined for new aircraft and hence incorporate a requirement for improvement in safety. Moreover, it was decided that there should be an order of magnitude difference between the upper and lower safety targets defining the ALARP region for any accident involving civilian aircraft. For accidents involving only military aircraft it was reasoned that the ALARP region should be two orders of magnitude, reflecting a greater tolerability of occupational risk within the armed forces.

2.2.3 *Overall Minsep Targets*

To derive safety targets in terms of minseps incidents from the mid-air collision safety targets it was necessary to estimate the likelihood of a minsep incident resulting in a collision. The likelihood was estimated by determining the ratio of minsep incidents to collisions from experience and also by modelling.

The data used to derive the minsep to collision ratio was taken from the set of AIRPROX reports that are filed whenever a near miss incident is reported. An AIRPROX report can be filed by air traffic controllers, in which case it is an AIRPROX (C) , or by aircrew, when it is known as an AIRPROX (P). AIRPROX's must be filed whenever the safety of an aircraft is thought to have been compromised by the close proximity of another. In controlled airspace, close proximity is defined as being within the prescribed minimum separation criteria - in most cases five miles horizontally from an aircraft and 1000 feet above or below.

The AIRPROX(C) and AIRPROX(P) data come from different sources. The CAA regularly publishes a digest of all AIRPROX(C) reports and these amount to eleven volumes [CAA 1990/96] covering the years 1990 to 1996. AIRPROX(P) incidents are kept in a database at Head Quarters of Military Air Traffic Organisation (HQ MATO) at RAF Uxbridge. Both sets of reports were used in calculating the ratio.

Confidence in the minsep to collision ratio estimate was gained by modelling the minsep airspace around an aircraft and determining the probability of the two aircraft colliding given that one is passing through the other's airspace in a straight line. The ratio was calculated by considering the relative areas presented during different flight paths through the airspace. The analysis showed a variation in the ratio for different aircraft combinations due to the sizes of the two aircraft modelled but was in reasonable agreement with the ratio calculated from the AIRPROX data. Moreover, the model was considered to be conservative in that it assumed the aircraft would fly 'blind' through the airspace, taking no avoiding action.

The model was used in conjunction with data from the United States Federal Aviation Administration (FAA) Near Mid Air Collision (NMAC) database in order to increase confidence in the ratio calculated from the AIRPROX data. NMAC incidents are reported by aircrew and tend to result from much 'closer' near misses (typically with 500 feet horizontally or vertically). Hence the FAA and the CAA data were not directly comparable. The model was used to scale up the ratio derived from the US data to compensate for the more stringent minimum separation criteria. Again, the results confirmed that the original estimate of the ratio was reasonable and conservative.

2.2.4 *Software Minsep Targets*

The final stage of the safety target setting process, apportioning part of the overall minsep target to ICCS software, involved three steps. The first of these steps was to derive a proportion of the overall minsep targets to apply to UKADGE military flying. The proportion used was equal to the proportion of military flying carried out under the control of UKADGE compared with the total amount of military flying hours.

The next step was to derive a proportion of the UKADGE minsep target to apply to the ATC equipment. The approach taken was to examine the AIRPROX(C) records to determine the relative proportions of the various causal factors. The data used were the summaries of each incident from the annual reports, each describing the events and main causes. Each incident summary was read and categorised, in a spreadsheet table, into the major causal types (e.g. ATC Human, ATC Equipment, Aircraft Human). The spreadsheet was then used to calculate the relative proportion of each the causal factors.

The results of the AIRPROX(C) cause analysis showed that 60% of minsep incidents involving (any) military aircraft were caused by aircrew (civil and military) and 30% were caused by air traffic controllers (civil and military). Of the remaining incidents, 9% were caused by a mix of aircrew, air traffic control and aircraft equipment and only 1% were caused by ATC equipment failure. The analysis highlighted that the human element of ATC is the major risk, and it was reasoned that equipment failure should be only a small proportion (less than or equal to 10%) of the causes of ATC related minsep incidents.

The final step in apportioning the overall minsep target to the ICCS software, that is deriving a proportion of the equipment failure minsep target for the ICCS software, involved estimating the relative contribution of software to the total number of failures of the UKADGE equipment. It was initially thought that data to support this estimate would be available from maintenance records for hardware and software. Although there was a reasonable record of software faults it was found that data on the actual numbers of serious failures, especially of hardware, and their consequence was unreliable and incomplete. After considering the data from the records and taking a consensus of operators' and system engineers' experiences, it was agreed that the best estimate of the contribution of software to serious failures was proportional to the perceived contribution of the software to the overall functionality. This contribution was perceived to be 30%.

3 Safety Claim

In order to make an argument for the safety of the ICCS, it was necessary to find evidence to support a claim that the software met the safety targets derived earlier. The following sub-sections describe the safety arguments used in the safety case and the evidence used to support the claims.

3.1 Safety Argument Overview

The first part of the main argument for the safety of the ICCS software relies on the claim that the UKADGE system as a whole has met its minsep safety target, and therefore that the ICCS software, which is a large part of UKADGE functionality, must also have met its safety targets. Clearly, this claim alone is not sufficient for a safety case since the aim is to assure the safety of the software for use in the future. Hence the second part of the main argument is a claim that although the software is continually being changed, the failure rate is reducing or at least being maintained.

A second, independent argument, was also made to support the main argument. This secondary argument relies on the claim that ICCS software satisfies its SIL requirements. This argument takes a 'white box' view of the software, and is intended to provide some confidence that the safety performance demonstrated for the main argument is indeed due to a measure of integrity within the software.

3.2 Safety Claim Evidence

3.2.1 Main Argument

The one advantage of developing a safety case retrospectively is that there are potentially many years experience of operational use in the real environment from which the adequate safety performance of a system can be claimed. However, the ability to extract the evidence necessary to support such a claim is greatly dependent on the quality of operational records.

Although the operational records kept by the RAF and by the ICCS software development team were directly relevant to the ICCS equipment and software, it was found that they contained little information about wider effects of the failures recorded (e.g. how it affected ATC at the time and whether an incident occurred). Hence, it was not possible to determine directly whether the equipment had met its safety targets.

The most reliable source of evidence to support the claim came from the AIRPROX (C) and AIRPROX (P) data. During the manual analysis of the AIRPROX (C) records (described previously) a special note was taken when a record described an incident involving military aircraft that were in radio contact with UKADGE control centre. The details of these incidents were of special interest, especially when equipment appeared to be a causal factor.

A similar analysis was conducted with the AIRPROX (P) records, although this was much easier since it was possible to search the database automatically for records with codes indicating that one of the aircraft involved was in radio contact with a UKADGE control centre. Once these incidents had been identified, copies of the full incident reports were printed for further examination. The database search was performed by the staff at HQ MATO, as described previously for the target setting.

For the second part of the main safety argument it was necessary to find evidence to support a claim that the current software maintenance program is not increasing the likelihood of dangerous failure. This was achieved by examining trends in the numbers of faults found and fixed, and also by determining a tolerable dangerous failure rate against which to monitor future performance of the software.

The analysis of fault trends made use of data contained in various records within the configuration tracking system. The problem reporting and fault fixing systems employed on the project were set up to ensure that reported problems were investigated and appropriate corrective action taken and recorded. However, it was never intended that these records of software changes would be used for any overall analysis. Hence, the analysis of the large numbers of records stored required a significant amount of pre-processing before any useful trend analysis could be carried out. One of the problems was that each record was a separate file, distributed throughout a number of directories and in some cases on physically separate computers. Moreover, although records were required to contain some standard information (e.g. dates), the use of free text meant that there were slight variations in how this information was actually recorded, making automatic searching very difficult. Another difficulty was that the software change records included records of changes made for reasons other than fault fixing (e.g. enhancement changes).

A software tool was written, by the safety team, to perform this pre-processing. One of the more specific pre-processing functions was to create a database of faults that would make it possible to determine the latency of a fault in terms of software baselines. This function was required to identify all the software change records that described the changes made to fix a particular fault and also to identify the problem records which related to this fault, in order to identify the point at which the fault was discovered. The dates were then traced and matched with dates of each software baseline issue.

The average dangerous failure rate was estimated for the last five years. A dangerous failure was defined to be one that caused a loss of, or impairment of, the ability to provide ATC even if the failure did not occur while an ATC service was being provided. It was reasoned that this failure rate must be tolerable, from a safety point of view, because during the last five years the software had been deemed to have met its minsep safety targets. This estimated failure rate provided a suitable basis for monitoring the safety performance of the software in the future.

3.2.2 Supporting Argument

There were two essential pre-requisites to the assessment of software against SIL requirement. The first was to identify the functional requirements of the software which are critical to safety (termed safety functional requirements) and the second was to define the assessment criteria.

The term safety functional requirements is used to encompass all requirements of the software which are associated with a safety related function. For example a requirement for a function may have associated with it other requirements such as timing, reliability, availability and capacity requirements. The hazard analysis identified the high level functions and failure modes that are safety related. The relevant high level functions were then mapped to individual requirements in the software specifications.

It was decided that assessment criteria of the software process would be taken from DO-178B [RTCA 1992] and it was argued that DO-178B Level C and DEF-STAN 00-56/2 SIL2 represent equivalent integrity requirements.

In order to facilitate the assessment it was decided to view the integrity of the software as being characterised by a set of properties, such as functional correctness, dependability, reliability and maintainability. This approach allowed the overall assessment to be broken down into more focused and manageable investigations. Each of the properties were further decomposed into sub-properties and each of these defined by a set of requirements which were derived from the relevant DO-178B objectives.

Although DO-178B describes the features of a process rather than the required properties of the software, it does give the objectives for each of the process activities. Hence it was possible to derive the basic requirements for each safety property by extracting the objectives from the relevant parts of the process description in DO-178B. For example, for the sub-property of functional correctness, 'software requirements are validated and verified', there are a number of relevant objectives in DO-178B sections 5.1 and 6.1 which cover the development (including validation) and the verification of high level software requirements.

For each of the sub-properties, the basic requirements were defined as described above and then these requirements were interpreted to make them specific to ICCS circumstances, for example by referring to particular specification documents. Evidence was presented in support of these specific requirements and a judgement made as to the extent to which the evidence supported the claim.

The collection of evidence of safety properties was carried out by initiating a number of individual investigations. Due to timescale and resource constraints it was not possible for the safety team to conduct each of these investigations independently. The investigations were carried out by providing members of the ICCS development team with questionnaires, developed by the safety team, to determine the extent to which various objectives of DO-178B had been addressed and met during the software development. The investigations covered traceability, test coverage, tool qualification, compilation and operating systems and procedures analysis.

4 Conclusions

The main conclusion to be drawn from the experiences described above is that it is possible to find enough evidence to develop a safety case, retrospectively for a large software system but not without some considerable effort.

It is quite likely that at the start of a retrospective safety case development no safety targets or safety requirements will exist. Evidence to support the setting of the safety targets is therefore likely to be required as well as evidence to justify that the targets have been met. In the case of ICCS, the general safety culture applicable to aircraft operations (civilian and military) meant that extensive databases of operational incidents were available. It was the existence of this data which made the setting of quantitative safety targets feasible. Throughout its life the software had been subjected to a disciplined configuration management and change control environment. Although the data in the change records had not been intended for trend analysis, meaningful trends were extracted from it.

The exercise of creating a retrospective safety case for ICCS has affected the safety culture of all concerned. The process of setting and justifying safety targets has resulted in monitors which will be used to judge the performance of the software in the future and further changes to enhance the confidence in safety continue to be identified.

References

[CAA 1990/96] *Aircraft Proximity Reports AIRPROX(C) - Controller Reported Volumes One to Eleven.* CAA, 1990 - 1996.

[HSE 1992] Health and Safety Executive: *Tolerability of Risk in Nuclear Power Stations.* HMSO Revised 1992.

[JSP318B 1996] JSP 318B: Regulation of Ministry of Defence Aircraft, 3rd Edition. Ministry of Defence, 1996.

[MOD 1996] Defence Standard 00-56/2: *Safety Management Requirements for Defence Systems, Issue 2.* Ministry of Defence, 1996.

[RTCA 1992] RTCA/DO-178B: *Software Considerations in Airborne Systems and Equipment Certification.* RTCA, 1992.

Acknowledgements

The authors are grateful for the co-operation and contributions of all who supported this work, and most notably RAF HQ 11/18 group, IoFS and SDC at RAF Bentley Priory, MATO at RAF Uxbridge, MOD(PE) SES24 and Adelard, who were the Independent Safety Auditor for the project.

Assuring Safety through Measurement – A Laudable Goal

Paul Goodman
TOK B Ltd (TBL)
Mead Lodge
Heathfield Lane
Chislehurst
Kent BR7 6AH

1 Introduction

Goals often fail to become reality. Can we in the safety critical systems industry allow the assurance of safety based on measurement to remain a goal?

The IT industry is a strange beast. We use language in totally incorrect and inappropriate ways that often leads to confusion among ourselves and certainly with our user or customer community. Let's start with a simple and trivial example.

I have, as in most families, the occasional conversation with my brother about the work each of us does. However, it was only recently that he raised the point that my answer to the question "What do you do?" was pretty meaningless. "I work in Software Metrics" is one answer I have given in the past. This, of course, means that I try to get centimetres and litres into software! Now, I know that more correct terms would be "measurement" but even that is only a part of what I do. Anyway, in our industry, rightly or wrongly, "Metrics" is an accepted term.

So now consider the title of this short paper, "Assuring Safety through Measurement – A Laudable Goal". I do not have too much of a problem with the first part but what about this "Laudable Goal" business? A goal is something we strive to attain but may not achieve, while laudable is simply another word for praiseworthy. Bear in mind that we are focussed on safety critical systems, systems for which the safety issue should be paramount. In the extreme, what we could be saying to the families of victims caught up in a disaster resulting from the failure of such a system is; "we were trying to make the system safe, for which we should be praised, but we didn't get that close to our goal!"

Of course, the words in the title of the paper are my words. Others in the industry may not use or even consider such an expression. But I have no evidence of the industry mindset being too different to that which I have expressed so far. In fact the situation gets worse!

2 Measurement in the Safety Critical Domain

Measurement in the safety critical area, based on fairly extensive research carried out last year by TBL on behalf of a client, appears no more advanced in this sector than in any other. There are exceptions, but the general pattern remains the same. If we take IT in general, based on other research by the UK Software Metrics Association, as many as 50% of organisations do not even measure effort expended at project level. To talk in terms of, for example, "measuring aspects of the software design in order to control complexity", which is one part of a Baldrige implementation, would be meaningless. So the safety critical domain faces the same problem as the rest of the IT industry, but more so! It is more important that we get it right.

One of the problems with measurement in the IT industry is that it is a big topic! One way of considering this is in terms of how measurement can be applied. For instance, one possible topic set would be:

- Management Metrics including product and process quality.
- Estimating processes and techniques.
- The use of Applied Design Metrics such as Information Flow metrics and McCabe metrics.
- Project Control.

Lots can be done. And it gets more complex because a measurement initiative within a project or an organisation, in reality, has to act as an agent of change, has to educate and train and has to enable genuine and rapid process improvement; because the products won't get better without people changing the way they do "stuff". Finally, the measurement programme will almost certainly have to develop and implement its own supporting infrastructure, both technical and in terms of personnel. The implementation of a measurement programme looking to institutionalise hard measures based on observed data is not a trivial undertaking.

Various authorities have quoted an estimated staff cost for a measurement programme of between 1 and 2% of the IT department overall staff cost. Typically, half to two thirds of this needs to be dedicated, full time resource. This implies that, being realistic, an IT department of 500 staff will need to allocate 5 to 10 person years effort per annum to the measurement initiative. Of this 2.5 to about 6 people will have to be assigned to staff a measurement function – full time.

At which point, most organisations will throw their hands up in horror and claim this is unreasonable. Bear in mind that most measurement initiatives fail! According to Rubin and Jones et al, and this suggestion is supported by our own observations in TBL, about 80% of measurement programmes fail within 2 years. Could it be that one of the major causes of this dreadfully high attrition rate is insufficient allocation of resources?

These figures and concerns relate to a measurement initiative within a project or single organisation. We have to consider a significant industry sector of many organisations, with its own distinct culture where the situation may be no better than the norm but where the price of system failure may be loss of life. If my premise is correct and the measurement culture in safety critical systems development and support is the same as or only a little better than in IT in general, we have been very lucky to date.

3 Problems with Introducing Measurement

I believe the problem we face is large and complex. It has a number of facets and I will attempt to summarise the issues as I see them. I do not believe this is an exhaustive set but it is a starting point. The issues are:

- Within safety critical system development and support, there is not an institutionalised, formal measurement culture within or across organisations in the industry.
- To introduce a measurement culture to an organisation is difficult and prone to failure. To change the culture of a whole industry sector may be impossible.

- Information depending on data does not come quickly or cheaply. We have to invest for the future.
- We cannot afford to do nothing. In fact we have to do something effective.

I would like to see a number of things happening. First, visible co-operation between organisations through the establishment of working parties to address measurement needs for safety critical systems at a national level. While some work has been done I believe we need a much higher profile initiative with the appropriate involvement of the industry, academia and government. This type of approach has worked well in the past. For example in the much simpler area of standards for functional sizing metrics where the usual bureaucratic process was short circuited during the development phase of the first standard in this area.

Second, bodies responsible for safety critical systems assurance insisting, at the organisational level at least, that suppliers have in place, can show and can be seen to be resourcing, strategic level measurement initiatives. As a minimum, these should span a three-year period.

Finally, individuals with any responsibility for the development and support of safety critical systems start to do things now. These do not have to be complex or far reaching but may simply be a start to obtaining defence in breadth and depth. A piece of work undertaken by TBL and Bournemouth University on behalf of a client took this approach.

4 A Simple Starting Point

We started from a position of wanting hard data. The requirements for data were based on taking a view of the "problem", i.e. the assurance of a system, that said it has four components:

- The system to be developed and the environment of the supplier.
- The non-technical environment of the supplier.
- The probability, risk and costs associated with failure of the system.
- The experience of the assurance team in making these kinds of judgements.

Each area was decomposed into a set of measurement requirements. Quickly it was realised that hard data would not be available from the majority of the suppliers for most of the identified measurement requirements. So we decided to accept informed, subjective opinion. Each measurement requirement was turned into a "directed question". For example:

What is the level of process maturity in the supplier organisation?

Became:

The supplier organisation exhibits high levels of process maturity.

The respondent then answered against a five point scale ranging from "Strongly Agree" to "Strongly Disagree" – and, yes, we had the debate about five and four point scales. The responses were weighted and added together to give an Assurance Ratio (AR). If the AR = 1 then this was a low risk project with a high level of assurance **based on the parameters of this assessment mechanism and the knowledge of the assessor(s).** As the AR tends to 0 then the risk rises and the level of assurance falls.

A very limited pilot of the questionnaire on one safety case, using a relatively inexperienced assessor, showed that the mechanism gave expected results. That is to say that the assessment scored the project as relatively low risk and safe and this was the experience of the client organisation that had been using the system for some time.

What we have is a crude, subjective assessment tool that can be used to support the assurance process. It was developed relatively quickly and is easy to implement. It is not the be all and end all, anything but. Further work in this area is underway and I believe we also need to address the other aspects of assurance, at least in part, through measurement.

5 A Final Few Words

Much more work needs to be done in this area, otherwise it will always remain a goal and I believe that the general public, our customers, would not consider that to be a laudable situation at all!

Assessing Safety Critical COTS Systems

Colin O'Halloran

L115, DERA Malvern, Worcs. WR14 3PS
c.ohalloran@eris.dera.gov.uk

Abstract. This paper gives a brief review of the problems of using Non-Developmental Items, NDI, (of which COTS products are a special case) in critical systems and approaches to dealing with them. The paper then concentrates on a particular technical approach which has been applied to a critical system comprising COTS, GOTS and legacy components from the point of view of demonstrating critical properties for certification. Finally prospects for extending the approach to NDI in open architecture systems are considered.

1 Introduction

There is a perception that COTS technology has the potential to deliver greater performance and value for money to customers especially in the defence world. The pace of commercial computing technology is so fast that some elements have a lifetime as little as two years, far less that the expected in-service life of any military system.

The customers of COTS products have little ability to control the development of commercial products. One problem for a customer is the cost of certification of a critical system. A related problem is the requirement for a stable environment in which COTS components can be substituted allowing the system to evolve without re-certification from scratch. If evolution is not addressed then the user is at risk of the whole system being rendered obsolete by the future unavailability of a single commercial component. The problem of assessing safety critical systems containing COTS products is a special case of assessing such systems with Non-Developmental Items, NDI. This general case is addressed in this paper.

2 The problem of Non-Developmental Items

2.1 Definitions

Commercial Off The Shelf (COTS) computing products are a special case of Non-Developmental Items (NDI) and are sometimes referred to as Commercially Developed Items (CDI). The acronym NDI also include Government Off The Shelf

(GOTS) products and denotes all those items which have been previously developed outside the system development.

Examples of NDI include hardware, software and other items available from commercial sources with or without modification to fulfill a specific task or role. A NDI is often a reuse of a software component, for example it could be a new application of civil software in a military environment. Other examples of NDI include: microprocessors; ASICs; FPGAs; operating systems and environments (such as X Windows); middleware, software which mediates between an operating system and application software; development tools and compilers, etc.

2.2 Issues

NDI may or may not be designed for a particular environment, for example a military environment. The environment in which NDI operate is important because the design of the item will often make unstated assumptions, clearly valid in one particular environment, which are invalid in a new environment for which the item was not designed to operate in. In general NDI contain functionality which is not required by a particular system, this is a commercial requirement since the more functionality present in NDI the wider the market for the item. Unfortunately the functionality present in NDI might not fulfill all the requirements of the system and the extra functionality present might contribute to a system hazard.

NDI may or may not have documentation available for a third party to review. For example full documentation for Microsoft products are not available even to Government organisations in the UK or USA even though such products are used in defence applications. There are advantages to using familiar Microsoft products, but there are disadvantages in terms of risk and cost. If documentation is unavailable this limits a certifiers ability to analyse or thoroughly test an item. In this case NDI must be treated as "black boxes". Even if documentation is present it might not accurately represent the actual product. Configuration management problems and time to market issues can result in an item which itself contains NDI from different sources at different times which the documentation does not record. If NDI do not have a configuration control process then there is a difficult issue of determining when changes to NDI have been made and what impact these changes have on critical requirements of a system.

Some NDI are modifiable, this allows them to be modified to meet the functional requirements of the system. When constructing a large system NDI will need to be modified in order to meet interface requirements. Modification of NDI addresses the problem eliminating unnecessary functionality. Unfortunately if NDI are modified then there is a certification problem. Modification can negate the established level of trust or assurance which has been established in NDI.

2.3 COTS characteristics

COTS products are a class of NDI with the following characteristics: they are readily available from commercial sources; they have general application; they are not designed for defence environments; documentation for COTS products are often not available; in particular access to detailed data is denied in general; configuration control is an issue; and COTS products are often not modifiable.

COTS products by their commercial nature have to be general applications, unless a specialised application has a high volume or value market. COTS product are therefore developed to be a "Jack of all trades". In particular a COTS product will contain functionality for compatibility with commercial products and functionality that is desirable for many applications. Unfortunately this additional functionality provides more potential traps for system level hazards to manifest themselves. In addition COTS products are not rigorously tested relative to their potentially critical importance.

COTS products are in general not intended for military environments. Most obviously they cannot withstand temperature extremes, mechanical or EM environments. More subtly they can make assumptions which are valid for commercial work or peace time work, but are invalid in high stress periods such as war time or an emergency. For example the Yorkshire Police COTS communication system worked reliably until it was stressed during the Hillsborough disaster, which was the critical time it was required to be available.

Documentation is a significant issue as far as COTS products are concerned. As mentioned before documentation for a COTS product might not be available, if it is available the cost of obtaining the documentation can be significant (a hidden cost of procuring COTS products). If documentation is available it might not accurately reflect the particular configuration of the COTS product. Any modification of a COTS product is difficult and ill advised without detailed accurate documentation. Configuration control is another significant issue, there may be hundreds of configurations of a COTS product with minor modification to the baseline occurring regularly. The commercial objective of selling a COTS product to as large a customer base as possible can lead to different configurations for different vendors.

2.4 Reuse

Reuse of computer products, especially software, is sometimes confusingly called COTS procurement. Reuse takes place in the Aerospace industry, but can crop up in defence procurement as Government Off The Shelf, GOTS, products integrated into larger systems. Unlike true COTS products, reused products tend to be specific applications which might or might not be suitable for another system. The reused product might need modification to satisfy a system's functional requirements or to remove extraneous functionality.

GOTS products tend to be easier to modify than COTS because of Government to Government agreements which allow direct access to source code or indirect modification by employment of a trusted third party. The documentation for GOTS products also tends to be more readily accessible and accurate than COTS products, although this is not universally true. Reuse of products developed within a particular company, which often occurs in the Aerospace industry, makes addressing these issues more straightforward.

Any certification for a reused product is application specific since reused products tend to be designed for a specific application. However some products are designed with reuse in mind which eases future certification.

2.5 Related issues

There are a number of related issues to the use of NDI in critical systems such as open architecture systems, technology refresh and language selection. Open architecture systems allow various vendors to supply components for a system and multiple sources for the same component. Slightly different products from different vendors can still meet the required form, fit and function. If there are multiple sources for a component then vendors will probably build in differences to attempt to gain a competitive edge.

In open architecture systems even completely different products can still provide the same required functionality. For example middleware, which mediates between application software and an operating system, can allow one software component to be replaced by a completely different software component as long as the required functionality for the system is present. The commonest example of different products providing the required functionality is when one microprocessor is replaced by another. In general an open architecture might potentially require multiple host computers which requires recompilation of the software to new target hosts. Open architecture systems have their own particular problems and opportunities with respect to safety assessment. For example the use of middleware presents the problem of assessing many components for what appears to be the same system, but middleware also provides a potential means of isolating components.

Many systems in the defence world are required to be state of the art or permit technology refresh. The lifetime of a military system is typically of the order of 30 years or more, in that time its technological superiority (and hence military superiority) will be eroded and overtaken. This is especially the case as far as electronic systems are concerned where computing systems can be obsolete within 2 years. If a system supports technology refresh then there is less risk from obsolescence allowing, for example, faulty components to be replaced by updated ones. This means a safety assessment and certification should not be viewed as a one off activity but an activity and cost which needs to be planned for during the lifetime of the system. Open system architectures provide an approach to dealing with technology refresh, therefore any assessment methods for open system architectures will also address technology refresh.

The choice of language can significantly impact the procurement and assessment of a system. For example C and C++ have many undesirable attributes, there are no validated compilers (in the sense of Ada compilers), and maintenance can be difficult. Even Ada contains language constructs which are non-predictable, that is the same source code can behave differently when compiled by different validated Ada compilers. These language issues have an impact on technology refresh even for open architecture systems and hence on safety assessments.

3 Dealing with Non-Developmental Items

There are basically three approaches to dealing with NDI in critical systems: ignore them; analyse and test them; or isolate them. A mixture of these approaches can be used in a particular system.

3.1 Ignore NDI

If NDI are ignored then they are essentially being treated as trusted components without supporting analysis, this is the least desirable approach. Any application software will have been developed in an NDI environment which can be of unknown quality. Any risk posed by NDI to a system has to be determined through testing alone. Trusting NDI means that no special consideration is given to potential effects of NDI on the safety critical functionality of a system.

There are benefits of trusting NDI such as lower development costs and less schedule risk to early development phases. The disadvantages are that there is no assurance that safety or mission critical or reliability requirements are met. Any configuration changes in NDI can also invalidate any test results. Trusting NDI is a high project risk approach which can lead to a system failing certification, or expensive retrospective analysis to produce certification evidence, or deploying a system with certification waivers and the consequent safety risk.

3.2 Analyse and test NDI

Analysis and testing of NDI is a desirable but expensive approach. For example to effectively perform an analysis of NDI documentation from the vendor might require it to be purchased, this can be expensive even when the NDI developer is only being paid to collect and collate the documentation. However analysis of NDI can be as rigorous as formal verification techniques applied to bespoke code, even though it inevitably will be more expensive to apply rigorous analysis retrospectively. For example the safety assessment of the NDI within the C130J avionics system using the Malpas analysis tools cost in the order of a million pounds after the purchase of the J version of the C130 (Hercules) aircraft. The safety assessment is nearing completion and has currently identified about 100 safety critical issues with the software which have been rectified.

In contrast Lockheed-Martin employed the same analysis techniques on a new software development for a mission computer using the SPARK examiner (in conjunction with Parnas tables generated during design) and they reported the use of the SPARK examiner led to cost savings in the software development. These savings mainly accrued from reduced costs for testing, without reducing the rigour of the testing, and faster turnaround of software after requirements changed. The post development analysis using Malpas also provided valuable visibility and control to the project office which in conjunction with the certification evidence produced, using goal directed analysis, is considered a success. However it is obviously more desirable to avoid this post analysis/certification situation if at all possible.

Analysis of NDI provides information about their structure and functionality which allows tests to be developed in the same manner that tests would be developed for bespoke software. Using a system hazard analysis and understanding the complete functionality of NDI through analysis allows hazard directed testing to verify the required safety properties of NDI. Information on the structure and functionality of NDI supports modification of NDI to reduce identified safety risks.

The benefits of this approach are that: assurance about NDI is assessed in the context of the system; it provides detailed knowledge about NDI; and steps can be taken to mitigate the potential causal factors of system hazards. The disadvantages of this approach are that: it is costly; detailed documentation might not be complete, accurate or reflect the version of NDI in use; there is an increase in the project risk (after all if safety problems are identified they will require rectification with the risk of cost and time overruns); NDI might not be modifiable; and an extensive test program will be required.

3.3 Isolate NDI

The third approach to dealing with NDI is to treat it as untrustworthy and ensure that differences will not impact safety critical requirements. This is the most cost effective approach. NDI are isolated from safety critical functionality through encapsulation mechanisms. The safety of the system is verified by demonstrating the effectiveness of the isolation and by both analysing and testing the safety critical properties. Of course this cannot be done for all NDI, for example faults in NDI used in the development of a system, such as compilers, cannot be isolated although their known faults can be mitigated.

Another problem is the integrity of data passed between isolated components which can impact the safety of a system. The key to solving this problem is to understand or precisely define the interfaces between components. In addition the data should be packaged using a defined format in order that it can be verified by application software at both ends and validation (using for example checksums) can take place. Unfortunately this might not be possible.

The benefits of this approach are that: it is cost effective if done early; it ensures that NDI cannot adversely impact safety; it minimizes impact of any future changes. The disadvantage is that it is not possible to isolate all NDI such as the microprocessor the software is running on, or the network upon which a distributed system operates.

4 Research results on NDI assessment

The Systems Assurance Group (SAG) at the Defence and Evaluation Research Agency (DERA) develops and applies leading edge technology for the assessment of critical systems. The process algebra CSP [1] and the model checker FDR [2] form key elements of the Group's capability. Recent SAG work has been the application of CSP and FDR on a critical distributed command and control system comprising COTS, GOTS, legacy and bespoke components.

4.1 A technical approach

The SAG in conjunction with Formal Systems in Oxford (the company which developed and market FDR) have developed a CSP constraint based approach to assessing such systems. The approach requires only partial information and is supported by the FDR model checker. The major concern of the assessment was the integration of a Unix based subsystem employing X Windows into a legacy command and control system. Because of certification problems the MOD customer approached the SAG to perform a system level assessment. It consisted of eight physically distinct but communicating subsystems. There was already sufficient information relating to the correctness and reliability of individual components obtained by detailed safety analysis, such as fault tree analysis. However no formal analysis of the interactions between subsystems had been performed to determine whether unsafe emergent behaviour could appear.

A system level hazard analysis had already been performed, this had identified a number of hazards which were grouped into a few hazard categories. Specific questions of interest were derived from the hazard categories and were carefully formulated so that they could be conveniently expressed as CSP refinement assertions.

From the point of view of formal analysis and model checking the most significant feature of this modelling task was the size and complexity of the system. In view of this it was originally believed that the system would not be tractable for model checking. The initial strategy was to model as much of the system as possible and then, when the state space exceeded feasible limits, proceed by making abstractions. However by the use of a new partial order method, previously developed for modelling security protocols [3], the system's properties were model checked quite easily.

The technical approach placed all the claims about a subsystem's behaviour in parallel synchronising over common inputs and outputs. The information about a subsystem's behaviour was derived from documentation and checked with the customer and supplier. Where documentation was incomplete or ambiguous the claim was weakened to ensure that the actual system behaviour was contained within the modelled behaviour. Although this is safe it leads to significantly more behaviours, or states, to be assessed. Separate validation took place to assess the accuracy of the modelled claims about a subsystem's behaviour.

The questions formulated from the system hazard analysis and expressed in CSP were used as specifications against which the modelled behaviour of the system, expressed as a conjunction of claims about subsystem behaviour, was checked. The refinement check showed that if the subsystems behaved as expected then the system would not manifest system level hazards.

4.2 Component failures

The assumption that each subsystem functions correctly is clearly improbable, however, because of the pessimistic approach to modelling the claims about a subsystem, the model can still be accurate even though the subsystem does not function perfectly. Unfortunately this is still not good enough and the failure modes of the modelled system have to be considered. In the same way that erroneous behaviour of computer systems can be explored by means of software fault injection, the failure modes of the modelled system can be assessed by injecting faults into the subsystems in the model.

The ability to inject errors into the model allowed the SAG to deal with random or systematic failures in a subsystem to determine their impact on the system, with respect to the safety properties identified from the system hazard analysis. The results of the analysis demonstrated scenarios under which a systematic or random error in certain subsystems could give rise to a system level hazard. These scenarios were presented to the customer and system integrators in order to determine whether the identified anomaly was an artifact of the pessimistic nature of the model, or was a genuine problem. In fact the problem arose because timing properties were not part of the model and could not be readily inferred from the system description.

Essentially the model revealed the possibility of a dangerous signal propagating through the system before any interlock could stop it. An independent model, which gave priorities to signals invoking interlocks over signals which invoke action by the system, demonstrated that the identified anomalies were due to race conditions. The CSP/FDR analysis technique is pessimistic, but counter examples to critical properties allow directed testing and analysis. Therefore the generated "dangerous" scenarios can be used to direct a testing programme to check that the race conditions are such that the system is safe.

The safety assessment showed that no single subsystem failure could give rise to a system hazard with the caveat that there was no underlying common mode failure which would manifest itself as two apparently independent failures. This caveat is the subject of further safety analysis. Checking the validity of the models is a separate assessment concern, but the modelling provides a means of focusing the assessment of the system information and the independent models were used to cross check each other. More technical detail about this work can be found in [4].

4.3 Extending the approach

Work within the SAG has been developing configurable encapsulations to add security to X Windows. Security arises from control of capabilities together with absence of bypass of the security mechanisms (that is the security controls). However the most significant problem for security is to add security to COTS products without reducing the functionality of a system to a level which makes it unusable. X Windows is particularly complex and, from a security viewpoint, vulnerable. This is because X was designed for use in an open community. The SAG have been investigating how security can be added without reducing the functionality of X to a level where applications cannot sensibly use it.

The SAG at DERA have designed an encapsulation mechanism for X whose configurations can be specified by a CSP description which can then be transformed into a Java implementation. The key to acceptable security is to identify capabilities, configure their control, and analyse the impact on security and functionality for a particular system. Either security and functional requirements can be simultaneously satisfied, or an informed trade-off between security and functionality can be made. A similar problem exists for safety and functionality.

Large information systems need to be flexible to deal with changing organisations and environments, COTS components with well defined Application Programming Interfaces (APIs) and an enforceable portability policy are essential to make the system flexible. Unfortunately even Open System components do not necessarily conform completely to Open System standards. However work within the Open Systems Software and the Systems Assurance Groups at DERA Malvern has been concerned with revealing APIs for languages such as C, Java and Ada. These APIs can then be encapsulated and policed to enforce critical properties and then assessed in a similar way to the approach described above.

The critical properties to be enforced on an open system constructed from COTS could be safety properties [5]. Indeed X windows is used in some weapon systems and can interact with embedded software onboard a piece of Ordnance. The encapsulation and high assurance policing techniques described in this paper would be applicable in situations where safety depended upon blocking the inadvertent actions of untrusted X clients.

Work on developing wrappers for encapsulation and investigating reflective architectural concepts to separate safety/security properties is being sponsored by the SAG at both Newcastle and Oxford University. Both rely on existing object structures, but work should be extendable to Open Systems with their implicit object structure. Combining this work with the assessment approach described in this paper provides an approach for assessing and isolating critical NDI in open system architectures.

5 Conclusion

The significance of the assessment approach described in this paper is that it is applicable to information systems, built from NDI, of a realistic size. It is applicable to open system architectures as well as networked systems where each component can be treated as a black box and a model built from system documentation and elicitation. The model will contain the behaviour of the actual system, this is a pessimistic approach which increases the state space significantly and report false anomalies, but it is safe and cost effective.

There are solutions to the state space problem, but it is not clear how generally applicable these solutions are in practice and is the subject of further research. The problem of "false-positives" turned out not to be significant and allowed the team constructing the safety case to focus on particular potential weaknesses. Another significant aspect of the work was the ability to "inject faults" into the system description to determine how robust it was. This allowed the SAG to confirm that the new introduced GOTS component was not safety critical and therefore reduced the potential cost of certification. This capability is very important in an era where the cost of certifying systems is becoming prohibitively expensive.

The future advocated approach to assessing critical open systems can be characterised as isolation of components in conjunction with analysis and test. First the system must be analysed as a whole to identify safety or security critical components and what critical properties they must have. Components can then be separately assessed or encapsulating software added to ensure critical property.

6 Acknowledgements

I am indebted to Mike Brown from the Naval Surface Warfare Center for raising assessment issues about NDI. I would also like to thank my DERA colleagues Irfan Zakiuddin, Nick Moffat and Peter Ryan.

7 References

[1] C. A. R. Hoare, *Communicating Sequential Processes*, Prentice-Hall International, 1985.

[2] A. W. Roscoe, *The Theory and Practice of Concurrency,* Prentice Hall series in computer science, 1998.

[3] A. W. Roscoe and M.H. Goldsmith, *The perfect "spy" for model-checking cryptoprotocols* , Proceeding of DIMACS workshop on the design and formal verification of crytographic protocols, 1997.

[4] I. Zakiuddin, N. Moffat, C. O'Halloran and P. Ryan, *Chasing events to certify a critical system*, (Draft) available from the author.

[5] C. O'Halloran, *Controlling the risk of COTS via Application Programming Interfaces*, proceedings IEE colloquium on COTS and Safety Critical Systems, Digest No: 97/013, 1997.

Systems Approach to Safety-related Systems

John Elliott

Defence Evaluation and Research Agency (DERA)
Systems and Software Engineering Centre, Malvern

Abstract

There is a consensus among safety professionals that 'safety is a system issue'. Despite this awareness, the relatively young discipline of safety engineering has yet to adopt a fully 'holistic' approach to systems development and subsequent assessment. This paper attempts to advance a 'true' systems approach to safety based on a modern systems engineering paradigm that provides a framework for future safety related systems development and assessment.

1 Background

Since the 1950's, systems engineering (SE) has advanced as a specialised, yet technology-independent, discipline primarily to tackle the management and control of increasing complexity within large scale engineering programmes. The generic SE goal is to ensure that a system's purpose is fully understood and satisfied throughout the product lifecycle. SE's emergence has partly been a result of the progressive development of system thinking approaches to help solve all kinds of problems through greater multi-disciplinary understanding.

System thinking achieves understanding about the relevant aspects of our complex world that comprise of dynamic, almost inseparable, (social, political and technical) systems. Although the consensus that 'safety is a systems issue' receives serious attention, see texts [Bahr 97, Leveson 95] and standards [IEC 98, MOD 95, MOD 96, MOD 97], modern systems thinking offers a way forward to understand and formalise the treatment of such system issues.

Over the last decade, the system engineering (SE) discipline has been making major advancements, assisted by many initiatives. Example initiatives include: Foresight [IEE 97]; INCOSE (International Council on System Engineering); the UK's EPSRC programmes (System Integration and SE networks of excellence); and current ISO standardisation programme, [ISO 98] on 'Systems Life Cycle Processes'.

The rigorous SE approach to system dependability and complexity management is highly desirable, especially in application areas where large losses may be involved. Naturally, SE is directly relevant to safety systems evolution where the key problem is achieving through-life safety assurance through a balanced satisfaction of system requirements over a range of mission related system properties, of which safety is only one such property (and constraint). SE performs this balance by adopting effective strategies in managing complexity and associated risks through examining wider 'holistic' issues. Overall, SE activities aim to achieve clearer: product requirements; product justification and evaluation criteria; interfaces between interoperable product systems; and links between design, certification and operational process systems.

This paper has been largely based on tracking the SE developments within a number of standardisation and research studies that have highlighted the opportunities for exploiting SE developments in the field of safety systems. Major standardisation framework developments have occurred in systems engineering, i.e. ISO 51288 [ISO 98], and in safety engineering, i.e. especially IEC 61508 [IEC 98]. It should be noted that, in the future, all system projects, safety related or not, will need to account for these 'standard' SE frameworks and the associated approach to the full system life cycle. The SE research initiatives are linked to the progressive development of a family of SE (including safety) reference models, addressing product and process characteristics. These models are aimed at improved understanding and measurement of concepts and terminology about systems, their properties and applicability.

Overall, this paper proposes that the safety systems community should pay more attention to the system-oriented frameworks, processes, principles and models being developed within different SE initiatives.

2 System Safety Engineering (SSE) Review

2.1 What is system safety engineering?

System safety engineering (SSE) is a compilation of engineering analysis and management practices that control dangerous situations. SSE achieves the following:

- Identifies the hazards in a system.
- Determines the underlying causes of these hazards.
- Develops safety engineering or management controls to either eliminate the hazards or mitigate their consequences.
- Verifies that the safety controls are adequate and in place.
- Monitors the effectiveness of safety controls during initial development or subsequent upgrades.

Using this safety management approach, SSE assures the safety of a system through effective design, but when is safe enough? Hence SSE also employs a set of risk assessment concepts to establish the risk of possible unsafe states with the aim of eliminating known safety risks or reducing them to an acceptable level, through review and iteration until declared as 'as low as reasonably practicable' (ALARP). These approaches are well documented in the standards IEC 61508 [IEC 98] and Def Stan 00-56 [MOD 96].

Leveson [Leveson 95] develops an understanding about the fundamentals of system safety starting with a model about accidents as having three levels: L1 - accident mechanisms, e.g. car skids, hits tree; L2 - conditions that trigger off mechanisms, e.g. driver untrained in skid control; L3 - constraints that allow condition to arise (the root cause of a accident), e.g. driver training policies excluded skid control. Leveson's analysis highlights the need for SSE to take a holistic view about accidents in the interest of avoiding them.

The common safety concepts affecting all accident levels and sequences are shown below in Figure 1. Although these safety concepts have derived from engineering experience, they have some foundation from system theories. SSE achieves safety through an effective process design (n.b. system engineers would call this a *'failure paradigm'* [Waring 96, Bignell 84]) complemented by a set of generic pragmatic safety design principles. Overall, SSE supports through-life safety design, its analysis and assessment.

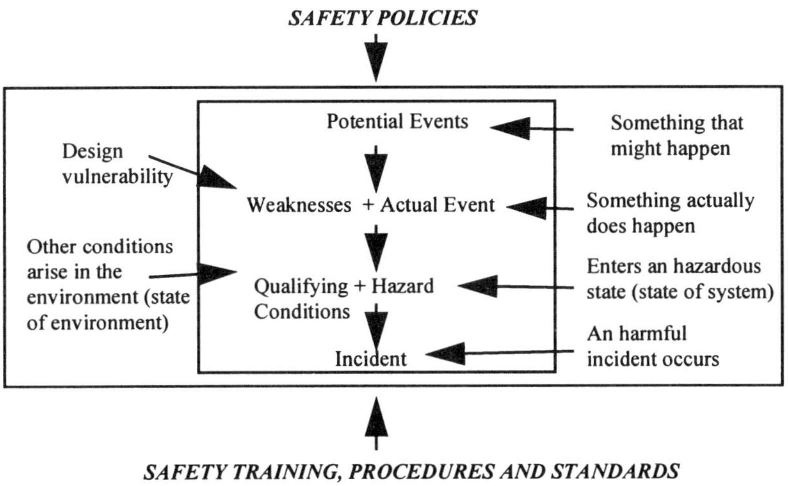

Figure 1 - Accident and Safety Concepts

2.2 Development of SSE frameworks

Of the various safety frameworks, the IEC 61508 standard [IEC 98] is a most suitable vehicle to illustrate some of the arguments in this paper.

IEC 61508 contains a safety management process applied to systems and components based on realisable technologies, with some supporting guidance on design criteria. The IEC emphasis is on the analysis and assessment of designs throughout the evolution of a *new* system development project. The use of these activities will influence the system design decisions but only as a part of any necessary risk reduction and related trade-off activities to comply with the safety integrity (risk-based) requirements. Thus IEC 61508 assumes: the existence of wider system engineering activities to manage and control cost-effective trade-off decision making; and adequate system understanding is achieved, crucial to hazard analysis and system requirements activities. Apart from the design analysis influence, the IEC safety lifecycle includes 'functional safety assessment' activities that enables the generation of an equivalent 'system safety case', based on the organised use of generated safety information, to be tailored to relevant certification regulations.

Overall, the IEC standard emphasises analysis and assessment throughout development, with limited guidance on safe system design. Also, as with many other standards, IEC 61508 is less clear on how to cope with common reuse, legacy and COTS issues. The IEC SSE framework is designed to apply to a range of electrical/electronic based technologies in use and providing some safety functionality. The IEC framework is very generic, a strength, so the standard applies to any system being implemented using one or more realisable technologies. There are a number of salient aspects of IEC 61508:

- The underpinning system model is based on a Total Systems concept that focuses on 'equipment under control', a controlling process system, external risk reduction features (e.g. physical guards etc) to mitigate effects, and safety related systems (e.g. warning, protection systems) based on electrical/ electronic/ programmable electronic and other technologies.
- The interwoven system and safety lifecycles, some of which are dependent on technologies in use, i.e. programmable electronic system, hardware and software levels.
- The interleaving between considering safety functions and safety risks (integrity) before completing a system and safety requirements specification to be used in planning and validation activities.

The IEC standard uses a generic *system lifecycle* within which a generic *safety lifecycle* applies; the system and safety lifecycle models use the same process structure. Two life cycles are shown in Figure 2. The rationale is that the system development lifecycle is so generic that its mapping onto any other local lifecycle can be easily achieved. The initial stages of the IEC lifecycle are technology independent embracing the different levels of requirements to architectural design representations.

The IEC's system and safety requirements specifications are divided into *functional* (for a safe and completed successful operational mission) and *integrity* (for such a mission but taken within a tolerable risk level) requirement components. The safety requirements depend on the iterative use of hazard and risk analyses of the proposed system, prior to identifying safety related functions and defined safety integrity levels; these analysis are supported by a 'tool-bag' of safety, often reliability, engineering techniques. The system and safety requirements analyses heavily depend on the 'quality/understanding' of the system description (models) of the system, its boundary and environment, and any assumptions. Subsequent use of such requirements information in system and safety planning and validation emphasises the need for clear system product definition and associated acceptance criteria.

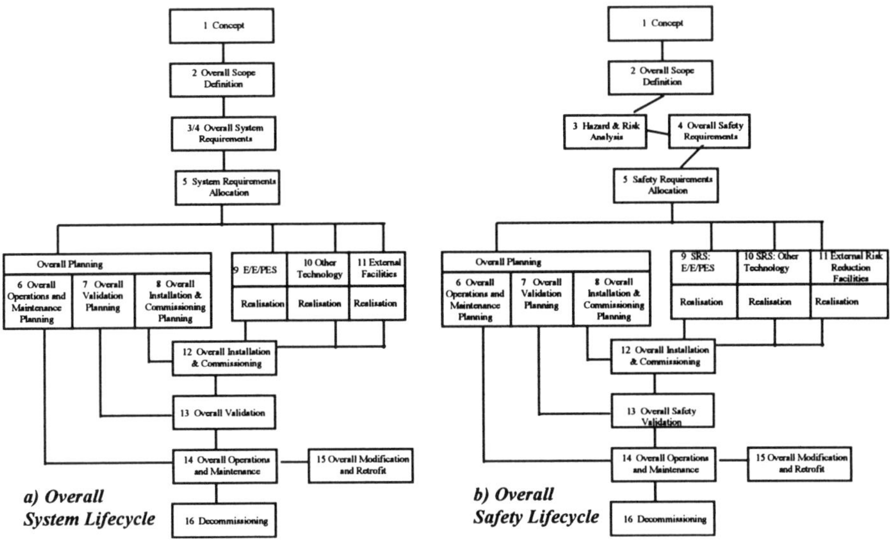

Figure 2 - IEC 61508 Lifecycle Models

2.3 SSE framework issues

In simple terms, there are four parts of SSE (that also relate to equivalent parts of system development in general) that need to be rigorous and continuously improved. These are:

- The derivation of *safety requirements* (i.e. analysis, capture and specification) from system understanding through models and analysis.
- The use of demonstrable and rigorous *design and verification* techniques enabling safety systems to be created with a low residual safety risk.
- The derivation of rigorous and traceable systematic approaches to *safety*

management, (i.e. integrating design, hazard and risk control).

- The *justification* of a system's 'safety fitness', linked to safety case development and assessment (i.e. safety information management, argumentation and uncertainty).

Overall the last decade, there has been improvements in these four areas, emphasising the role of computer based safety related sub-systems. However, there is further scope for improving system-wide processes by adopting a more formal systemic, not just systematic, system engineering approach to safety engineering. The overall need is for wider holistic vision throughout the SSE approach.

3 Systems Approach

3.1 Origins/motivations of systems age

The origin of an systems age [Skyttner 96, Leveson 95] arose from a need to explain combined 'emergent' phenomenon that could not be adequately explained using individual sciences and associated theories. The early interest in systems was started by a fascination of analogies between biological and physical phenomena in the early 1900s. This was in contrast to an earlier machine age where reductionism (i.e. investigation of the parts) was thought to be the way to understand and determine behaviours of combined sets of components; this assumes that the sum of the parts equals the whole, now realised to be only valid in the most simple cases and where established physical, not abstract, laws apply.

System scientists tried to develop a 'general system theory' (GST) that is an attempt to create a 'science of sciences' using holistic system concepts, to be independent of all other physical and social sciences. GST was intended to explain phenomena that cannot be addressed by the individual sciences, e.g. phenomena that arises when two component explanations are unable to explain the impact of higher order emergent observations. Subsequently, 'general system*s* theory' (GS*s*T) emerged where the emphasis is on understanding 'systems' as a separate and useful science for thinking about the world, to be used to help explain multi-disciplinary challenges across many problem domains.

As the system age developed from the 1940s, all system thinking developments tended to identify variants about system theories, models and methodologies. The main variants about the implementation of system thinking, founded on GSsT concepts, is shown in Figure 3. These variants are based on particular viewpoints of a system theory linked to its purpose and application. *System science* is a response to GST and GSsT developments to determine a science and underpinning theory that is geared to the application of models and methodologies, not concepts for academic excellence. *System analysis* is an aid to pragmatic and rational management problem solving where problems were structured prior to their solution. This variant is closely linked to the operational research/management sciences discipline that also involved the use of cybernetics to recognise that socio-technical behaviour could be viewed using biological analogies. This emphasised control, decision making and communication via adaptive-feedback behaviours, all embodied in Beer's Viable Systems Model [Beer 66].

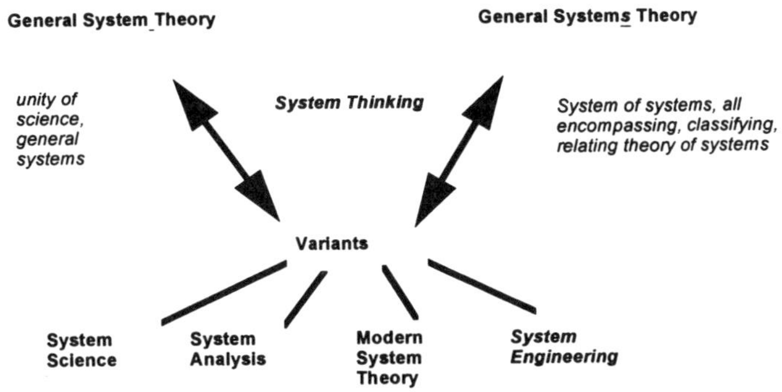

Figure 3 - System Thinking Variants

Modern system theory has emerged based on the ultimate acceptance that the human being is a part of most systems, often with a key influence on the outcome; this led to the development of 'hard' and 'soft' system ideas[1], [Checkland 84]. Checkland founded a major system classification scheme, a system typology, as in Figure 4.

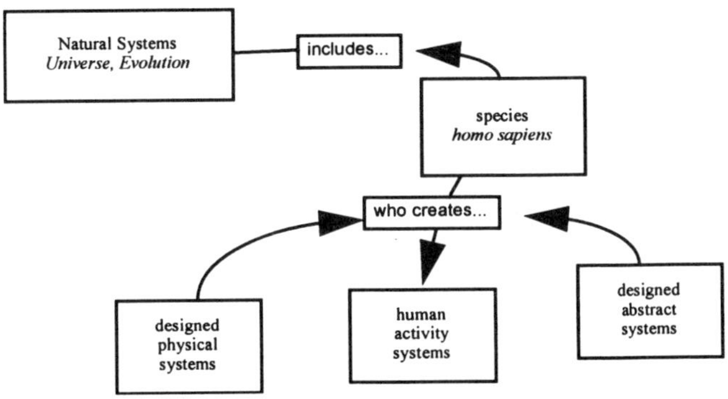

Figure 4 - Checkland's System Typology

1 Hard systems (e.g. engineered products) are different from soft systems (e.g. human activity systems) in that they are more predictable, more quantifiable, less (minimal) dependent on human variability, generally more purposeful, less subjected to political and social influences and more focused on technological aspects.

This typology separates out the fundamental types of systems: the natural biological systems (e.g. weather); engineering systems (e.g. computers); abstract systems (e.g. programming languages); and human activity systems (e.g. project teams and organisations). All real world systems use one of more of these types. The separation of the soft 'human activity systems' recognised that problem solving often involved incomplete understanding largely due to social and political influences, not just technological. This modern system thinking led to the development of various system methodologies based on different paradigms. Acknowledgement of the subjective nature of systems and their boundaries was based on the interpretative nature of how humans view what is relevant given their purpose, i.e. what problem that they need to be solved.

The last, but not least, development from GSsT is that of *Systems Engineering (SE)*, that was originally intended to be a 'hard' methodology to build systems in a realistically deterministic manner, yet accounting for the 'soft' issues involved. The soft issues arise when defining the problem to be solved through interactions with users and other stakeholder interests, and when defining the process being used by a project teams to solve the problem. Hence soft issues are crucial to 'defining the right product' prior to 'developing the product right'.

The early SE progress since the 1950's was directed towards system management as a controlling and co-ordinating activity, to be based on the system concepts to be applied. These concepts have been expressed using various SE development models. Not only has the subject of SE progressed to consider the hard and soft issues, but it has enhanced the linkages between theories, models, and methodologies, already evident in the development of the system science, system analysis and modern system variants. SE is discussed in more detail in Section 4.

3.2 System foundations

This section outlines the concepts of particular importance to safety systems. The essence of the system approach is to tackle problems, by externally viewing reality as a whole system with a definable purpose but within its interacting environment; it is essential that the properties of the environment are understood. Figures 5 and 6 show the system concepts; further details are found in [Skyttner 95, Kramer 77].

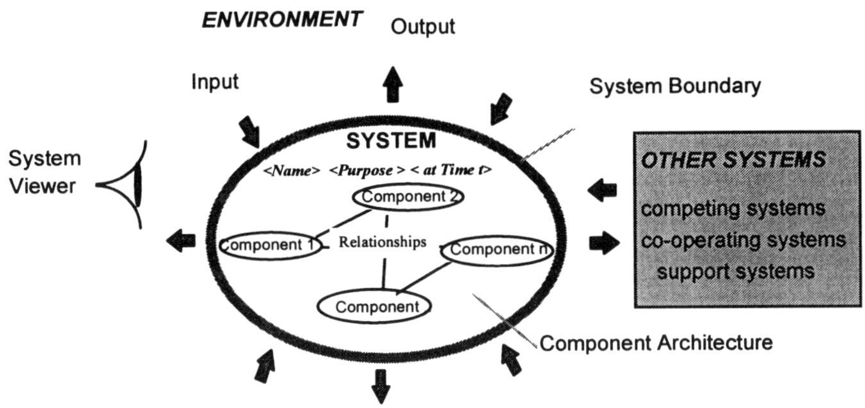

Figure 5 - Systems, Components and Environments

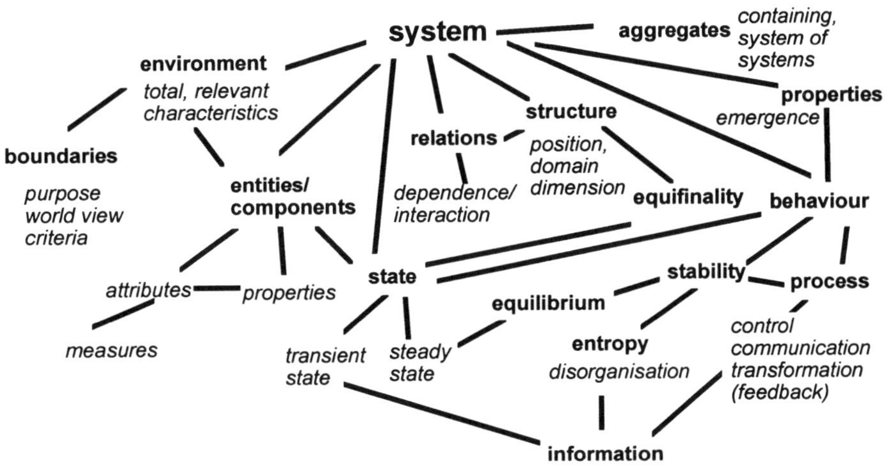

Figure 6 - Systems and their Concepts

The salient aspects of systems that are of particular relevance to safety systems are:

Emergence - This is where the whole system behaviour is a result of the interaction of its parts, e.g. wheels, pedals, frames, etc join together to form a machine that moves, a bicycle. The whole purpose of defining requirements is to define desired emergent properties. Other undesired emergence properties may arise such as hazards arising from combination of failures, the cause of many accidents.

Hierarchies - This is where the relationships between systems are defined to accommodate abstraction and viewpoint concepts. This leads to describing systems as being contained within larger systems, forming 'systems of systems'. Communication of information about the relationships between higher or lower level of systems in any such hierarchy is essential for understanding and confirmation purposes. These ideas are fundamental to safety requirements, hazard analysis and safety case models.

Complexity - This is a kernel issue to be controlled but is affected by the number of components, the number of interactions but also by the degree of uncertainty about the system or its environment. Controlling complexity is often through the use of abstraction to aid understanding. However in critical systems, the abstract models of system architectures should enable, where appropriate, the re-design and risk reduction measures to minimise complexity.

Control and communication - These aspects are central to all systems remaining stable and exhibiting the desired emergent properties. Control requires information from a controlled system through some feedback cycle, enabling the system to adjust the controlled system's behaviour. In safety systems, the various type of functions such as diagnosis, monitoring, protection and control are obvious examples.

Structure and behaviour - System structures identify the internal organisation of some of its functions whereas the behaviour is the result of the functions responding to environmental interactions. This is why safety behaviours are our prime concern (the whole emergent affect of functions) through the design and implementation of defined functions.

Open and close systems - Closed systems are self-sufficient (e.g. an endless clock) where open systems depend on their interactions with the environment. Most systems have some degree of openness. In safety systems, all systems are open, and thus we need to understand the feedback and learning cycles that arise; these cycles may enhance safety through effective hazard control or may lead to uncertain and undecided actions.

4 SE Emergence and Developments

4.1 Overview of SE

There have been many attempts at defining SE, each emphasising a particular aspect. Here are one example definition:

'A set of activities which control the overall design, implementation and integration of a complex set of interacting components or systems to the needs of all users.' [Stevens 98]

Different definitions vary the scope and nature of SE that generally embraces technical, management and business activities undertaken through the lifetime of a system. It is usually more helpful to consider the nature of SE. This is highlighted in Figure 7, showing the essence of the SE approach to solving problems by creating system solutions.

The SE emphasis is: to seek goals and decision criteria early in problem solving; to apply a creative and pragmatic approach to generating alternative solution; to develop and use models of systems in the analysis and evaluation of solutions; and to adopt mechanisms for controlling the product development to satisfy the needs.

SE originated to address difficult problems within systems development. Such system difficulties are where there appeared to be high costs, too many functions and parts, too many interactions and (unknown) side effects, human-machine interaction issues, and too many environmental unknowns. In order to cope with

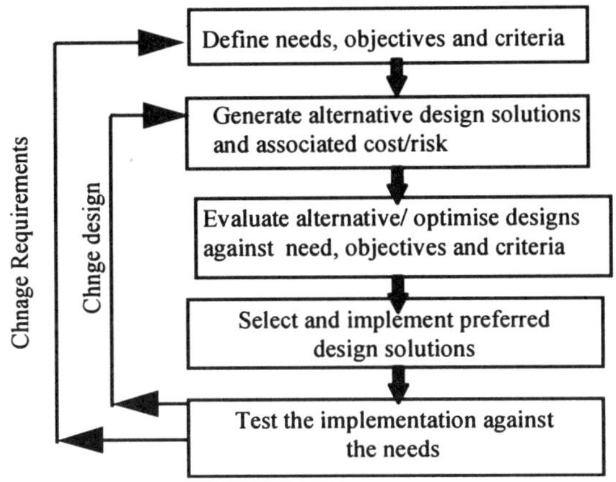

Figure 7 -Essence of System Engineering

such system difficulties, SE controls any complexity by adopting the use of abstraction techniques to determine different viewpoints at different levels in the hierarchy of systems (i.e. business, project and technical products). Figure 8 emphasises the relationship between complexity, viewpoints and abstraction.

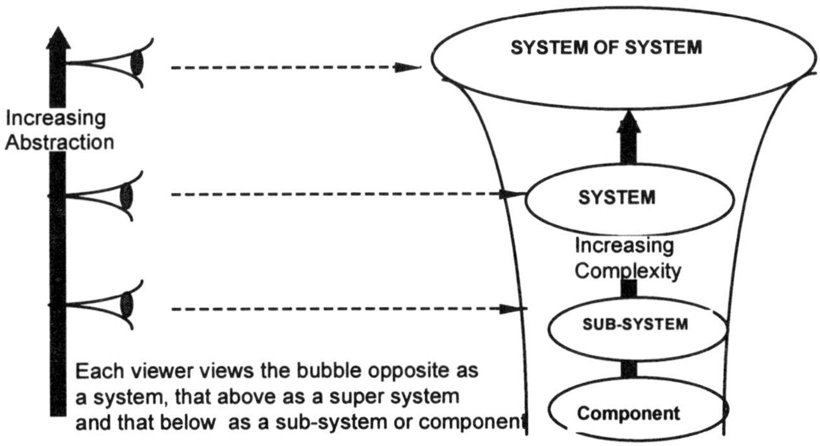

Figure 8 -Abstraction, Complexity and Viewpoints

These different views are critical to successful engineering as they allow the system to be viewed by procurers, users, designers, implementers and others. The highest level view is where a *'business capability'* goal is defined and developed into a business-level system architecture, in which contained systems inter-operate; this is sometimes referred to as the *'system of system'* level that describes the overall containing business system of interest. Each system of systems part is likely to be complex and the subject of a separate project activity. The lowest level is the component level, that may not be further subdivided, and this may be a new, COTS or legacy system. Note that these systems or components may be any system types [Checkland 84] including human activity systems.

Apart from using abstraction and hierarchies, SE tackles system difficulties by considering:

An holistic view. SE takes a multi-disciplinary 'whole' view of the system and its environment (e.g. using integrated project teams) to enhance the understanding of the business needs, any user's requirements and key success criteria. SE needs to be whole-life oriented to ensure the 'right product' is identified, whilst allowing for real world changes in requirements within a realistic evolutionary flexible lifecycle approach.

An user-oriented view. SE focuses on user needs and specifies required behaviours (what the system does) prior to defining system structures (how it does it). However, such structures are developed to be technology independent to address broad and future users' need. SE does not become unnecessarily concerned with implementation technologies too early.

System separation. SE focuses on separate systems of interest, possibly at different enterprise, project and technical levels, separating 'hard' (technical) and 'soft' (human) systems. Of importance is the establishment of well defined interfaces between these systems.

A balanced view. SE needs to control and satisfy any systems assurance requirements, through a product engineering management process. SE will manage the balances through rigorous technical trade-off activities considering the optimisation of whole system risk and design change implications.

Creativity. SE processes need to allow adequate 'time and cost' space for creativity but within given constraints. SE largely depends on creative activities such as achieving systemic understanding, and generating and evaluation alternatives.

Infrastructural Issues. These are issues are affect the efficiency and effectiveness of the SE approach: the composition of the team structure; the design of the SE process; the use of tools; and the associated management of system information.

Overall, 'pure' system engineering consists of defining rigorous system processes to include systemic, systematic, model-based, technology-independent and whole-life approaches. The system engineering approach is particularly relevant where any of the following apply: large-scale systems are involved; where system dependence is high; and where financial and technology commitments are significant and have long lasting effects.

4.2 Introduction - SE Reference Models

Section 4.1 has described the key attributes of the SE approach, the basis of characterising the SE process and the result system product. One way of advancing the SE subject has been to develop SE reference models. The aim of a reference model is to describe the terminology and understanding in terms of suitable concepts and their relationships. Any such model may be used as a generic 'reference' framework from which to generate variants tailored to business organisations, project and technical areas. Furthermore, these models may be sub-divided into sub-models that each focus on specific viewpoints and perspectives capturing the properties of SE processes and products, as well as related system understanding concepts. Hence such sub-models may act as a reference for key technical areas, of which safety may be one such area; safety processes are a part of the SE processes and safety functions as a part of the SE product.

Overall, SE reference models may address the process of creating systems or the characteristics of the evolving system product. To date, a number of initiatives have led to the development of different types of reference models embracing systemic, process and product aspects. These will be summarised in Section 4.5.

4.3 SE Process Reference Model Description

This section concentrates on the SE process view of relevance to the development of ISO 15288, SE lifecycle process. This process view has been captured as a generic *system engineering process reference model* (SEPRM) framework. The SEPRM has been designed to further the evolution of SE standards and guidance for global standards or developing tailored organisational practices. The SEPRM is essentially a wide ranging process model that has been deliberately designed to explain all SE situations, for all abstraction and viewpoint levels. This description introduces the main elements of this SEPRM, further details are in [DERA 98, Stevens 98].

The scope of SE is addressed by the SEPRM in two ways by:

- Defining a full system process lifecycle supporting different levels of abstraction and associated viewpoints, and major process interfaces.
- Defining a hierarchy of system products.
- Relating lifecycle processes to each product.

Process levels. SEPRM has been developed to encapsulate the varying levels and perspectives about SE through a system's lifetime. The SEPRM describes the business enterprise, project and SE technical levels of activities, and their associations. Figure 9 shows how these levels are simply formed as well as being able to capture complex relationships in multi-organisations, projects and SE processes/products.

This complexity may arise by following a 'supply chain', distinguishing between the acquirer (e.g. procure and customer), developer (e.g. systems integrator and architect, sub-system/component supplier), operator (e.g. operations manager and system users) and supporter (e.g. maintainer) interests.

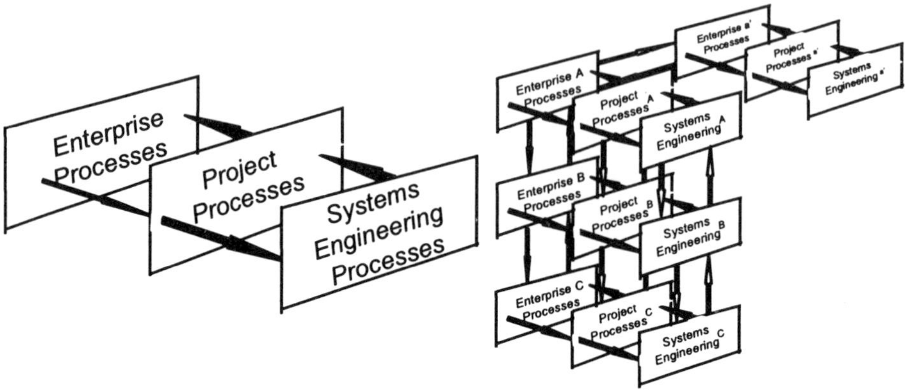

Figure 9 -Levels of System Engineering Influence

Full system lifecycle. Fig 10 shows how these interests vary throughout the system life cycle that address the acquisition, development, operation and support stages of a system's life as well as including various customer-supplier enterprise aspects. It should be noted that each part of this life cycle is a potentially separate process system linked to a separate enterprise activity; the links between each process systems is the transfer of products and associated ownership. This process and product view will be revisited in Section 4.4.

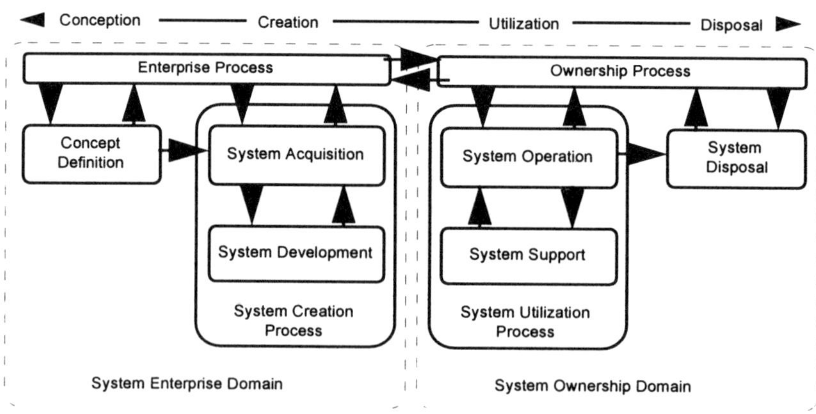

Figure 10 -Full System Lifecycle

It should be noted that SE's prime activity is to create a systems within a enterprise and project context prior to the transfer into the system utilising process within a ownership context. Thus modern SE takes a holistic view of the lifecycle and considers the impact of the enterprise process's created system prior to its full operational use and support.

System creation. Figure 11 shows the sub-processes within the system creation process.

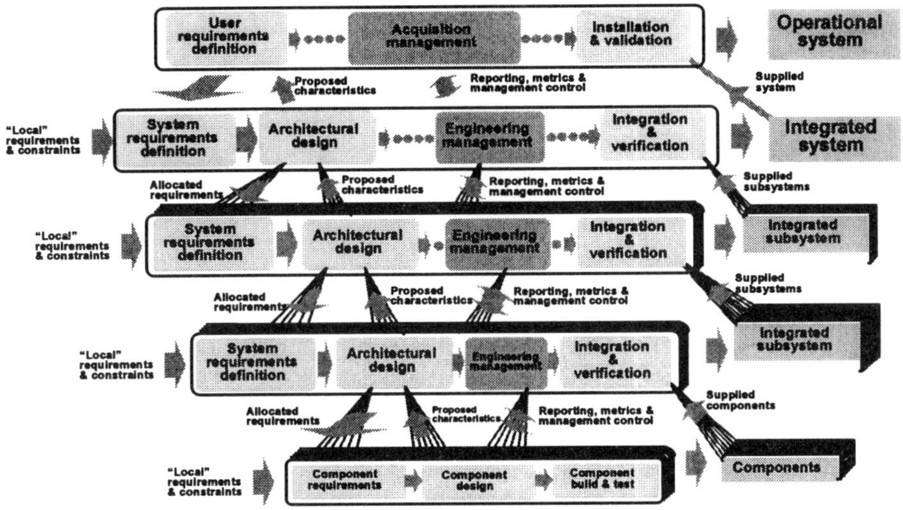

Figure 11 - System Creation Lifecycle

Process and project interfaces. Figure 11 assumes an early concept stage has been undertaken to define and fully understand the business need and that the output will be a validated system to be moved into operations, through real-world trials until the system is finally approved and accepted for operational service. This is where a whole or part business capability idea is developed, depending on the level of abstraction and viewpoint (e.g. system of systems project, systems or sub-system projects). It should be noted here that the overall acquisition management sits above all system development, undertaken by a lead enterprise role, who will ensure that all system or component level projects have effective and working interfaces. This is highlighted in Figure 12 that shows more details about the appropriate interfaces.

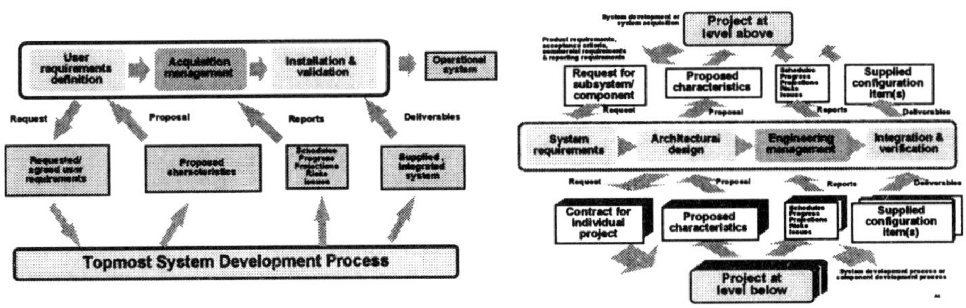

Figure 12 - Acquisition and System Projects

Major SEPRM processes for safety systems. These SEPRM processes below are particularly relevant to the safety system activities.

> *'User requirements'* is where the 'right product' is decided with criteria for judging the fitness of the resultant product. Using the systems approach within early concept studies and an user requirements stage, the business need is determined and the requirements are derived in a rigorous manner, following good requirements management practices.
>
> *'Acquisition management'* ensures that an acceptable system is created within the system project constraints to meet defined needs and requirements. For higher levels of system of systems, this activity includes programme management, where many complex projects may be required to produce inter-operating systems, to fulfil a fully planned business capability target.
>
> *'Engineering management'* includes all technical management activities necessary to ensure the resultant product complies with its user requirements as well as possess a demonstrable and adequate quality profile. This includes any kind of technical risk management, including risks associated with safety and other system properties. Also, this process ensures that all design trade-offs are based on rational analysis and can be shown to meet an agreed and prioritised set of design criteria embracing all aspects of the required quality profile. For large scale projects, this process is sometimes referred to as product engineering management in support of higher level acquisition led programme management.
>
> *'System architectural design'* is fundamental in the SEPRM as the system progresses towards a technology dependent set of sub-systems or components. SE would have completed its role as technical co-ordinator if the architectures set out well defined and robust interfaces, having analysed the combined effect of the functions to generate desirable behaviours only, i.e. useful emergent properties.
>
> *'Various hand-over and system delivery activities'* (including analysis, trials, assessment and justification) are necessary when approving a system as 'fit for its intended operational use' or when considering follow-up modification and support. Following system verification in a development and test environment, there are two levels of acceptance involved: provisional system acceptance based on evidence from trial in the real operating environment but under test conditions; and final acceptance based on a period of detailed operational use of the system in the real world.

4.4 Revisiting SE Process or Products

All systems behave (produce an output) in response to stimuli (an input). The behaviour depends on the functioning of all the interacting components. The relationships between process and product systems within the context of SE, as in Figure 10, is vitally important to understand.

Essentially Figure 10 shows six process systems (concept, acquisition, development, operations, support and disposal). Each process system generates product(s) to pass into other process systems; for example: the requirement and criteria pass from the acquirer's process system to the developer's process system; the accepted and certified product originated from the developer's process system passes through acquisition into different modes of operational use (linked to levels of system integration and acceptance).

Of course, the complex product systems have many components as hierarchies, but the same process system relationships apply to each position within the product hierarchy. The only difference is that the process system will be subdivided to generate the various sub-systems prior to system integration, all of which is described by the SEPRM.

Ultimately, the design of process systems must adequately cater for the dynamics of human activity systems, see the earlier description of Checkland's typology. These human systems actually apply the processes and make judgements based on decision processes, information, training and competence. Hence there is 'human friendly' process design job to be done, an issue only known too well by the social sciences. So the systems approach to process design needs to account for the human, not just treated as another add-on 'hard' function, but as an unpredictable element within a 'soft' social process. Hence, following on from the earlier Checkland system typology, development process systems are largely human process systems but supported by, and interoperate with, support product systems (i.e. software tools) and procedural systems (i.e. local practices). Similarly, operational process systems may involve inter-operability between human process systems and other (e.g. legacy) product systems.

In safety system processes, the need for a consistent systemic approach to system safety processes demands that humans and their reliability are vital components within the processes of system design and system operation.

4.5 Other SE Reference Models

The SEPRM is a process model that captures the scope and essential process of creating and using systems. There are three other SE reference models to briefly mention and to refer the reader to other texts. These models capture the properties of SE processes and products, as well as wider system understanding concepts. These models are:

System understanding reference model (SURM) [DERA 99] - under development, to be based on a range of system thinking styles and problem solving scenarios. This model is used to determine problem solving strategies based on system properties and types. This systemic model promotes the evolution of through-life 'customer satisfaction', requirements understanding and associated system measurement (addressing 'fitness of purpose', system properties, complexity and uncertainty) for all levels of system abstraction. One use of the SURM will be to devise 'fitness' system arguments analogous to those created within safety cases.

SE Meta-process Model, (SEMM), [Oliver 97] - an alternative useful SE process model to contrast with SEPRM. SEMM captures a combined management and technical process model for each SE stage. Uniform and generic core SE technical processes are defined, each repeated through different stages of SE (i.e. the product lifecycle). Also the SEMM offers simple viewpoints about behaviour and structure hierarchies and project management.

SE product model - Generic Reference Model (GRM), [Hitchins 92] - a part of an interesting and stimulating attempt to develop a theoretically founded SE methodology. The GRM[2] is a model of the system product, not process, that has two parts:*function* model (what the system does) and*form* model (how the system does it); both are regarded as an internal view of the system. An external*behavioural* model view of the systems product has been under development. The GRM encapsulates various system design properties and can be used to guide system design, to check designs for completeness, to help in the diagnosis of system deficiency and to evaluate systems. Also the GRM can be used to develop a means of studying emergent properties to examine the properties of an actual system model. These GRMs are derived from a series of SE principles that forms an understanding about systems based on modern system theories.

These other SE reference models offer some potential for developing a composite understanding of safety systems, as processes and products. One of the major benefits offered by these SE reference models, the SURM and GRM in particular, concerns support for making trade-off decisions about safety functions and emergent behaviour but in a wider context of satisfying a mission comprising of many other non-safety properties, (e.g. functionality, security, reliability, availability, cost and so on). There is an emphasis on system and requirements *understanding* supported by a *measurement* framework, see [Fenton 97], from which to judge overall system fitness, i.e. a 'system case', at all levels of system abstraction from *macro* system of systems to *micro* components levels. These reference models identify some key areas to support the creation of safer systems.

5 Systems Approach to Safety Engineering

5.1 General

This section outlines how the systems approach embodied in modern SE can be used to further enhance safety systems development and assessment. The following areas are discussed to indicate the potential direction for SE and SSE to complement each other:

- Integrate the use of SE (ISO 15288 based on SEPRM) and SSE (IEC 61508) standardised frameworks.
- Design a product system evaluation and certification scheme.
- Design a SE and SSE process and tool environment.
- Demonstrate the 'fitness' of a safety related product system.

---

[2]The GRM (function) addresses: Mission, Viability and Resources. The GRM (form) addresses: Structure, Influence and Potential. The prospective GRM (behaviour) addresses: Characteristics of responsiveness and Stability.

5.2 Integrating SE (ISO 15288) and SSE (IEC 61508)

The integration of the two IEC and ISO (SEPRM based) approaches amounts to: developing the interfaces between the two process systems; identifying overlaps and missing links; and enhancing the guidance offered in either process system. A comparison of each process system is given in the table below.

Systemic area	**IEC 61508**	**ISO 15288 (SEPRM)**
General systemic approach	•Main focus is on technical systems, of any scale. •Sets out 'user' and system requirements as context without further guidance. •Uses 'total systems' model and thorough analytic approach to safety. •Needs system definition to do safety studies to determine boundaries, no guidance how to do so. •Focus on safety function within given system architecture, to be the subject of analysis. •Measure of achievement (i.e. safety) emphasises dependence on quantitative reliability approach.	•Takes broad approach for enterprise, project and technical SE process levels. •Stresses 'holistic' approach, geared to large complex and dependable systems. •Allows for both hard and soft systems. •Offers guidance how to apply system thinking approach to identify interacting systems/components and their properties. •All system architectural guidance available for analysis of all properties. •Measure of achievement (i.e. all requirements) dependent on qualitative and quantitative approach geared to system mission effectiveness.
System understanding and requirements	•Part of 'system concept', to understand the EUC and its environment (physical, social, legislative, political) although no guidance. •Referred to as a system requirement, a given element. •Safety requirements derived using hazard and risk analysis that influences the appropriate safety functions to be included in system functions.	•Use of system thinking techniques to describe the environmental elements. •Stems from enterprise business needs to introduce a new or changed system allowing for the environment and its changes. •Main trust of SE is to ensure a well founded user requirement statement based on sound macro-system analysis. •All properties such as functionality, maintainability, costs, etc all part of the whole-life treatment of systems.
Technical management	•Use of risk assessment best practice to determine the safety integrity level, a reliability measure, but derived using reasoned arguments about the impact of identified hazards. •Assumes that safety risk reduction occurs within a system review cycle to consider design changes to improve safety properties towards an acceptable risk class.	•A uniform approach to all risk management is proposed, including composite controls. Hence the holistic approach to risk allows a balanced view about safety and its value. •Defines 'engineering management' process to handle and balance all kinds of technical risks, for different layers of system (business capability to components). •Guides the trade-off process supporting the risk reduction activities, but for all properties and their prioritisation.

Systemic area	IEC 61508	ISO 15288 (SEPRM)
System measurements and acceptance judgement	•Detailed account of safety verification and (independent) function safety assessment. •Same approach to whole system as with technology based sub-systems. Basis of Safety Case.	•Main strength to measure the degree of system compliance. •Sets the reference point/criteria for judging compliance. •Engineering management needs to justify decisions for the whole system and all its properties, and hence the basis of a 'System Case'. So it supports safety case ideas but again from a holistic viewpoint.
Technology realisation	•System and safety requirements provide the overarching technology independent architecture. Limited guidance available. •Details biased towards technology safety lifecycle process and design specifics. •Includes technology-specific system components safety lifecycles and design specifics.	•Totally technology independent, will point to technology guidance on systems not just safety. •The requirements derive from the system architecture, supported by SE. •The safety requirements for sub-systems and components should be derivable from the architectural details.

As expected, the SE view (SEPRM based ISO 15288) is broadly supported by system thinking approaches. However this summary analysis shows how the SE system understanding and different system measurement areas appear to be quite complementary to the needs of SSE (using IEC 61508). Furthermore, the other SE reference models, outlined in Section 4.5, will add further support to SSE. Overall SE offers further support on a balanced 'holistic' approach to define and measure individual and composite product system properties, beyond safety. The treatment of unwanted emergent behavioural properties is vitally important and is an area to enhance the traditional SSE hazard and risk techniques.

5.3 Designing System Certification scheme

The strength of the SE approach is in deriving clear goals and measurement criteria. Hence it is reasonable to consider applying the SE and systemic approaches to designing and analysing the complex world of safety assessment and certification. This is an area where there is a mix of factors with which to wrestle: objective and subjective statements; observable facts, uncertainties and incompleteness; technical and social acceptability; legislation frameworks; enterprise relationships; evaluation criteria; evaluation processes (evidence generator); certification process (fitness judgement generator).

A typical simplified evaluation system is shown in Figure 13. This system is discussed in [Rae 95], and the subject of a current EUROSCOPE certification initiative[3].

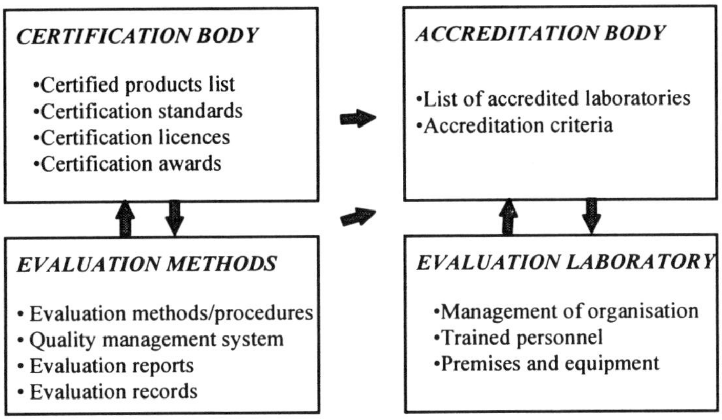

Figure 13 - Outline Certification and Evaluation Scheme

This is, in reality, a complex and important process system that needs to be fully understood in terms of a certification environment and evaluation system framework, together with its products and processes, and its inputs (especially evaluation criteria and user requirements) and outputs. Whilst subjectivity is unavoidable, the quest in safety evaluation should be to minimise its impact on variability, e.g. competence schemes is one current approach but new evaluation methods and criteria are other avenues to explore. Also the above evaluation scheme need not be divorced from the developer's and acquirer's design processes, so a 'design for evaluation' view is important to maintain a whole-life systemic view, as suggested by Bishop and Bloomfield in [Redmill 98]. Whilst outside the scope of this introductory paper, the systems approach to certification can provide a useful review to improve the current approach to certification thinking.

5.4 Designing a SE Process and Tool Environment

As introduced in Section 4.3, SE may be used to design and understand the subtle interactions between process and product systems, and recognising that each are made up of component systems, often of different types. For example, development process systems are a combination of human activity systems (project teams, different stakeholder teams), procedural systems (technical, management and business lifecycle practices) and development tools/environments (physical and logical). This represents a typical mix of hard and soft systems where a broader problem solving approach may be helpful, particularly as some risk-based

[3] EUROSCOPE is a network of companies pursuing excellence in systems and software assessment and certification.

measure of process effectiveness (i.e. the integrity and usability of the resultant products) is needed. This is analogous to Hessami's proposals in [Redmill 98] for a holistic approach to risk management. One of the big problems, especially in software based systems engineering is the dependence on effective processes and where there is a dearth of real metrics. A systems approach to measurement may uncover the more significant risk areas in process designs, that may led to risk reduction measures to the SSE process, as advocated within safety system product developments.

5.5 SE and Safety (or System) Cases

The subject of safety cases would benefit from a systemic approach to broadly explore the measurement issues, especially that of the system product. Fundamentally such cases contain information that sets out justifications (as arguments, evidence, assumption and inferences) about safety claims. In fact, a safety case is another abstraction and viewpoint about the system product. Hence such product information provides an important basis upon which major judgements and decision are made about safety 'fitness for purpose'. A systems view of how the safety case contributes to the overall decision for the system to be used in operations, under test or full operational conditions, would ensure the efficient and effective role of safety cases. Naturally we need to worry about the safety case process, a lifecycle, again as proposed by Bishop and Bloomfield in [Redmill 98], that generates the safety case product. This safety case process will direct and control supporting concepts and techniques for the creation, organisation, manipulation and management of safety information.

Our real concern should be to produce systems that fully satisfy business user needs as well as being 'safe'. This is where system engineering comes into its own, by directing and co-ordinating activities to create a well argued and balanced system solution that is operationally useful, yet adequately safe. Together with SE support for system measurement, the vision is a composite 'system case' that will allow a balanced and thorough justification and assessment of all system properties, of which a safety case will be one view. Already safety case concepts have been extended to apply within the use of systems reliability, particularly in defence systems.

6 Conclusion

The overall conclusion from this paper's analysis is that there are clear benefits in bringing together system engineering and safety related initiatives through a co-ordinated approach within standardisation and research programmes. Current SE initiatives have significant inertia to support safety influences being discussed in ISO, INCOSE and similar forums. The synergy between IEC 61508 and ISO 15288 is plain to see. The advancement of SE reference models provides a key to understanding, measurement and integration of SE with SSE.

The common interests between SE and SSE in system thinking, understanding and measurement are all of paramount importance and there is more to be done to formalise the systems approach, especially concerning its applicability to safety systems.

In fact, although safety related systems is a specialised topic, the fruits from safety related process research could, and should, be applied to support the development of system engineering and the management of other system properties, such as security and reliability.

Disclaimers

The views expressed in this paper are entirely those of the author and do not represent the views, policy or understanding of any other person or official body. The author would welcome any exchange on knowledge and concepts contained in this paper. He can be contacted at jjelliott@dera.gov.uk.

References

[Bahr 97] Bahr N J, *System Safety and Risk Assessment: A Practical Approach*, Taylor & Francis, 1997

[Beer 66] Beer S, *Decision and Control - The Meaning of Operational Research and Management Cybernetics,* John Wiley and Sons, 1996

[Bignell 84] Bignell V, Fortune J, *Understanding System Failures*, Manchester University Press 1984

[Checkland 84] Checkland P, *Systems Thinking, Systems Practice,* John Wiley and Sons, 1984

[DERA 98] DERA *Systems Engineering Reference Model* - DERA Internal Report

[DERA 99] DERA - *Systems Understanding Reference Model* - Internal Report (to be completed)

[Fenton 97] Fenton N, Pfleeger S L - *Software Metrics*, (Second Edition), Thomson Computer Press 1997

[Hitchins 92] Hitchins D, *Putting Systems to Work*, John Wiley and Sons, 1992

[IEC 98] IEC 61508, *Functional safety of electrical/electronic/programmable electronic safety-related systems*, International Electrotechnical Commission, 1998

[IEE 97] Foresight Report - *Building Integrated Systems*, IEE, 1997

[ISO 98] DISISO 15288 (Draft) - *Systems Life Cycle Processes*

[Kramer 77] Kramer N, de Smit J, *System Thinking - Concepts and Notions*, Leiden 1977

[Leveson 95] Leveson N G, *Safeware - System Safety and Computers*, Addison Wesley 1995

[MOD 95] Def Stan 00-58, *A guideline for HAZOP studies on systems which include a programmable electronic system*, Ministry of Defence, 1995

[MOD 96] Def Stan 00-56, *Safety management requirements for defence systems*, Ministry of Defence, 1996

[MOD 97] Def Stan 00-55, *Requirements for safety related software in defence equipment,* Ministry of Defence, 1997

[Oliver 97] Oliver D, Kelliher TP, Keegan JG, *Engineering Complex Systems with Objects and Models*, McGrawHill, 1997

[Rae 95] Rae A, Robert P, Hausen H - *Software Evaluation for Certification - Principles, Practices and Legal Liability,* McGrawHill 1995

[Redmill 98] Redmill F, Anderson T - *Industrial Perspective of Safety Critical Systems,* SSS'98 Proceedings, Springer-Verlag, 1998

[Skyttner 96] Skyttner L, *General Systems Theory - An Introduction*, Macmillian Press, 1996

[Stevens 98] Stevens R, Brook P, Jackson K, Arnold S - *Systems Engineering - Coping with Complexity*, Prentice Hall 1998

[Waring 96] Waring A, *Practical System Thinking*, Thomson Business Press, 1996

The Safety Analysis Case in the São Paulo Metro

João Batista Camargo Junior, Jorge Rady de Almeida Júnior
Digital System and Computer Engineering Department
Polytechnic School – University of São Paulo
São Paulo, Brazil
e-mails: jbcjunio@pcs.usp.br, jrady@pcs.usp.br

Abstract

The purpose of this paper is to present the experience acquired with the studies about Safety Analysis, realised for "Companhia do Metropolitano de São Paulo", the company that manager the Metro system of São Paulo City, focusing the signalling and control systems of the metro net.

The current São Paulo City Metro net and its future expansions are described, highlighting some technical and operational features.

The main body of the paper describes the Safety Analysis Methodology, under continuous evolution and improvement, according to the new implemented microprocessor technologies.

A primordial aspect is the extreme importance of the independency between Suppliers, Operators and the Safety Analysis Group.

This paper covers several stages of this methodology, including the Safety Analysis of analogical and digital hardware modules, software modules and operational procedures involved.

1. Introduction

The safety control system in a mass transport vehicle is an essential issue for the users and for the Operator's credibility. In order to assure the system safety, it should be emphasised the usage of techniques that allow prevention of accidents or, ultimately, the minimization of their consequences.

Accidents generally lead to harmful consequences for human lives. Material damage, cancellation of missions, and environmental consequences should be also considered. When a loss is considered very serious, it generally complains for the efforts and resources not invested in the prevention of those accidents [1].

The related activities to the safety should already begin with the conception of the system and they should have sequence along its project, production, test and operation and maintenance. The project should emphasize, since its beginning, the related aspects to the safety, avoiding later increases, and so adding other elements for that purpose. The safety should be seen as a whole, that is to say, it is not enough just to guarantee the safety only over some components of a large system.

It is of primordial importance the establishment of a Safety Analysis Methodology, seeking the verification of all the related aspects to the safety in the operation of the Metro control systems.

The control systems used in Metro transport are having an increasing automation, in function of the need of performance improvement and increase of transport capacity, which implies, in consequence, in larger complexity of those systems. That automation has been accomplished mainly through microprocessor circuits, due to their wide advantage in what refers to the control and supervision capacity.

It is worth to stand out, however, that the application of those microprocessor systems in the automation of the Metro transport, demands a careful verification of safety's aspects involved in each application [2].

An existent aspect in the control systems that use computers refers to the employment of software modules, responsible directly for functions that implement safety requirements. The non-attendance to such requirements can lead the system to unsafe states, that is, states where there is the possibility of occurrence of accidents.

In chapter 2 of this paper it is presented, in a summarized way, the current configuration of the São Paulo Metro System, also including expansion plans. It also stands out in this chapter the technological evolution that is being implemented in these lines.

In chapter 3 it is described the Safety Analysis Methodology that is being used and improved since the beginning of the implementation of São Paulo Metro System.

In chapter 4 it is presented the conclusions and some current recommendations due to the Safety Analysis Methodology, as well as due to many works of safety analysis accomplished in several projects for the Company of the Metro of São Paulo.

2. Configuration of the São Paulo Metro System

The studies for the implementation of the São Paulo Metro System began in the middle of the decade 60. The construction began properly in 1968, and the commercial operation began in 1974 [3].

Now three lines compose the São Paulo Metro, totalizing 44 km of extension (most underground), including 41 stations. The Metro works from of 5:00 a.m. to 12:00 a.m. The average distance among stations is approximately 1.1 km.

From the 41 component stations of the current Metro net, there are 4 stations with intercity bus terminals, 16 with urban bus terminals, besides 5 stations with rail integration.

The potential number of users of the São Paulo Metro System is approximately 18 million people, that is to say, the population of Great São Paulo. However, the ratio of the line extension to the number of users is now in the range of 2.5 km of line per million of inhabitants. There are several cities on the world that have this number varying between 20 and 50, indicating clearly that there is still need of great expansion of the current São Paulo Metro net.

The transported passenger number is in the range of 2.5 millions/workday, totalizing 700 million passengers transported a year.

The headway used in the rush hour is around 100 seconds and the fleet is composed by 588 cars, being 528 used in the rush hours. Each train is formed by 6 cars.

The three component lines of the São Paulo Metro net are:

- **Line 1 - Blue**: It was the first to enter in commercial operation, having the track interlock system controlled through vital relays. The trains are controlled by means of microprocessor digital technology with duplicated architecture, having its outputs compared by means of fail safe circuits. On board computer programs are an active part of this subsystem. The maximum speed allowed in that line is 100 km/h.
  - **North extension**: This extension of line 1 is being implemented now. It has track interlock system implemented with the use of microprocessor technology, with no use of vital relays in any part of the system. The architecture of this subsystem is triplicated, having its outputs voted by means of fail safe circuits.
- **Line 2 - Green**: It was the first track interlocking system implemented in a commercial line in Brazil with the use of microprocessor technology, with no use of vital relays in any part of the system. The architecture of this subsystem is triplicated, having its outputs voted by means of fail safe circuits. The trains are controlled also through microprocessor digital technology, with duplicated architecture, whose outputs are compared by fail safe circuits. Computer programs are an active part of both subsystems: track and trains. The maximum speed allowed in that line is 75 km/h.
- **Line 3 - Red**: It has the track interlock system controlled through vital relays. The trains are controlled through microprocessor digital technology, with duplicated architecture, whose outputs are compared by means of fail safe circuits. The yard interlocking system of Itaquera was the first implemented in Brazil, using microprocessor technology in a non commercial line, with no use of vital relays in any part of the system. The architecture of this subsystem is triplicated, having the outputs voted by means of fail safe circuits. The maximum speed allowed in that line is 87 km/h.

The expansion plans of the Metro net for next years are:

- Modernization and adaptation of already existent rail lines - about 50 km; and
- Construction of new aerial and underground lines - about 50 km.

The São Paulo Metro basic fare is about one American dollar for free use for any of the lines, besides integration fares that exist with buses and trains.

The figure 1 presents the configuration of those lines.

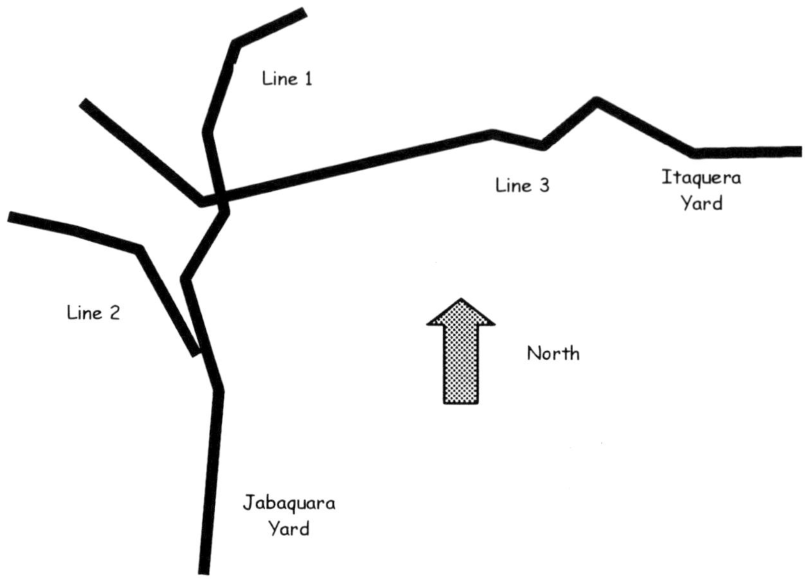

Figure 1 - Configuration of the São Paulo Metro Lines

3. Safety Analysis Methodology

The Safety Analysis Methodology presented in this paper has been performed to the microprocessor digitals circuits and fail-safe circuits, which are used in the control systems of some lines of São Paulo Metro. This methodology consists of a coherent group of activities, with the objective of improving the warranty of the computational system safety, applied in critical areas. It is worth to stand out that this methodology is being improved and developed since 1982, when occurred a technological exchange between Battelle Columbus Laboratories, of Columbus, Ohio, USA, and the Polytechnic School of the University of São Paulo, through an agreement coordinated by the Company of the Metro of São Paulo.

It is important to highlight that the evaluation of the safety of a critical system should be framed in a global culture of safety involving the Operator, the Supplier, as well as the Independent Safety Analysis Group. The concern with the safety must be present in all the phases of the life cycle of the system, including also the training aspects and the involved professionals' motivation.

The harmony and clear understanding of the concept of safety among the members of this structure really allow assuring a larger level of safety for the system. In this sense, the Safety's Paradigm adopted by the Company of the Metro of São Paulo is formed by the independence of roles among the several entities, highlighting the importance of the complete independence of the Safety Analysis Group in relation to the Suppliers of the system, as presented in figure 2.

Considering the adopted methodology, Safety is defined as the probability of the system not to reach an unsafe state, that is to say, when the system is exposed to an accident. In this sense, the unsafe state is considered a subset of the failure state of the system.

Figure 2 – Relationship between the entities involved in Safety Analysis

Considering the adopted methodology, Safety is defined as the probability of the system not to reach an unsafe state, that is to say, when the system is exposed to an accident. In this sense, the unsafe state is considered a subset of the failure state of the system.

The failure state is reached when the system carries out its functions incorrectly. The elementary origin of this failure state is denominated fault, which represents an imperfection inside of a certain component, which could be a hardware or software one. The manifestation of this fault is reflected in the contamination of the internal system information, being characterized as an error. The manifestation of an error to the external environment represents a failure state, which can be safe or unsafe.

Figure 3 presents the logical relationship between these concepts.

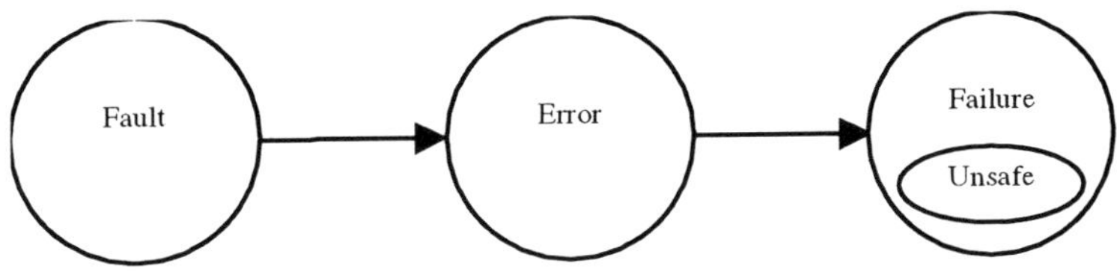

Figure 3 –Relationship between the concepts fault - error - failure

This Safety Analysis Methodology is composed by the following stages:

- Stage 1: Determination of the Safety General Requirements - SGR
- Stage 2: Mapping of the Safety General Requirements
- Stage 3: Analysis of Safety Responsible Modules
- Stage 4: Safety Final Evaluation

Based on non-technical aspects in Company of Metro of São Paulo, nowadays the safety analysis process is conducted after the technical project is finished. In spite of this, the Company of Metro of São Paulo intends to perform the new safety analysis process at the same time of the signalling system project.

Stage 1: Determination of Safety General Requirements - SGR

One of the fundamental aspects in the evaluation of the systems that carry out safety functions is the determination of Safety General Requirements (SGR). These requirements are generated from the elaboration of a General Description of the System and of its Architecture. Safety General Requirements try to identify unsafe situations.

In the identification of the unsafe situations, it is fundamental the accomplishment of periodic meetings with the operator of the system, trying to extract the largest amount of information regarding to its operation. In the specific case of the São Paulo Metro, it was considered as unsafe situations the derailment, the collision and the running over of maintenance teams. Based on the determination of the unsafe situations, it can be determined the Safety General Requirements, using techniques such as Fault Tree (Top-Down) and Failure Mode and Effect Analysis - FMEA (Bottom-Up). We intend to incorporate still in this stage the Hazard and Operability Technique - HAZOP to computational systems [4,5,6], as another available tool for the determination of the Safety General Requirements. A support tool was already developed to build and evaluate Fault Trees.

At the end of this stage it is of fundamental importance a critical evaluation of these Safety General Requirements by the Operator of the system.

Stage 2: Mapping of Safety General Requirements

This activity takes place on the several modules of the system, being determined those that have the attribution to assure the safety in the operation and maintenance of the system. These modules can be hardware components, software components or even operational procedures. As consequence can be determined the Safety Specific Requirements - SSR.

For the appropriate implementation of this stage, it is necessary the elaboration of a Functional Description of the System (Hardware, Software and Operational Procedures). Besides aiding in the determination of Safety Specific Requirements, it also supplies necessary subsidies for the next stage (stage 3) of the methodology, in which a Detailed Description of the component modules of the system is accomplished.

In this stage the techniques of Fault Tree and FMEA also can be used.

Stage 3: Analysis of the Safety Responsible Modules

In this stage are analyzed the modules that implement safety functions. There are specific evaluation methods following described.

Stage 3.1: Hardware Module Analysis

Initially the hardware modules are classified as fail safe implementation and redundant implementation, and they are analyzed according to the following methods:

Stage 3.1.1: Fail Safe Modules Analysis

A fail safe module must respect a fail safe degree. We can say that a normal state (state without failures or improper inputs) is considered fail safe with degree **n**, when the reached state is not unsafe, even with the occurrence of **n** events of sort **detectable** failures and/or **detectable** improper inputs. The specification of the fail safe degree in São Paulo Metro is equal "1".

This stage consists of the following activities:

- Elaboration of a Detailed Functional Description of the module in normal operation, that is to say, without the existence of components failures or improper inputs;
- If necessary the Safety Specific Requirements are more detailed;
- Realization of FMEA; and
- Realization of Improper Inputs Analysis (for example: noise, source oscillation, electromagnetic interference, etc.).

Stage 3.1.2: Redundant Modules Analysis

This stage consists of the following activities:

- Analysis of each redundant module, including the same activities described for the fail safe modules;
- Analysis of the independence of each channel of the redundant module, regarding to its implementation, power supplies and input signs, trying to identify Common Failures Modes - CFM.

Stage 3.2: Software Modules Analysis

It is worth to highlight that all the routines that constitute the built-in software in safety's equipment are appraised. These routines can be directly responsible or not for a safety aspect. This procedure refers to the fact of the high dependability degree among the software modules. The activities that are part of this analysis are now described.

Stage 3.2.1: Documentation

It consists on the preparation of the necessary documentation for each routine and variables/constants of the software. Referring to the software routines the following Forms are elaborated:

- Presentation Form, containing the name and function of the routine, implementation language, file to which it belongs, called and caller routines;
- Safety Requirements Form;
- Functional Description Form;
- Structured Diagram or Fluxogram Form;

- Data Flow Diagram Form;
- Test Cases Form, embracing input values and expected results;
- Conclusion Form, filled at the end of the stage 3.2.3, described further.

For each variable/constant in the software, the following Forms are elaborated:

- Variable Description Form, containing the name, type and dimension of the variable, file of declaration, routines that read it and routines that modify it, beside the functional description of the variable;
- Constant Description Form, containing the name, type and value of the constant, file of declaration, routines that read it and the functional description of the constant.

Stage 3.2.2: Formal Code Inspection

It consists of a formal inspection of the source code through the application of a specific checklist for each language. Such check-list also tries to embrace some eventual critical points mentioned by the supplier of the compiler itself. Some of the typical topics verified in that formal inspection are:

- Code termination: termination of functions, ending of hardware and software interruptions, loops termination. . .
- Control Flow: verification of complex constructions, correspondence between enable and disable of interrupts commands, . . .

Stage 3.2.3: Meetings of Walkthrough

Walkthrough is a formal meeting between the analysts involved in the work with the objective to complete the analysis of the system routines. In those meetings it is named for each routine a defender, that has as basic purpose the explanation of the functional aspects of the routine. The other participants of the meeting have as objective the enquiring on the logic of the routine, trying to identify situations that can lead the system to unsafe conditions.

Among the participants of this meeting it is fundamental the presence of at least one professional involved with the analysis of the hardware, trying to aid in the identification of failure situations connected to hardware failures. Those meetings should have the maximum duration of 2 hours, in a way to maintain the constant attention of the participants on the aspects of system safety. One of the participants of the walkthrough should be named as the reporter of the meeting.

To demonstrate the behaviour of the routine, its defender accomplishes several Cases of Test on that routine. For the accomplishment of these tests some fundamental criterion can be mentioned, for example:

- Logical covering, whose objective is to test the functional aspects of the routine;
- Verification of improper and border values;

- Verification of independent paths, whose objective is to test the structural aspects of the routine;
- Error guessing – test cases based on the experience and on the distrust of the technical team on some aspects of the system;
- Verification of its Completeness [7,8], in which are verified some basic criterion that are divided in categories: Input/Output Variables, States, Input/Output Predicates, Input/Output Relationship and Transitions.

A Specification Evaluation Environment is being developed relatively to the completeness of some criterion, using specification representations by means of Adaptive Statecharts [9,10]. This environment is to be used also in the system level, besides being able to aid in the determination of new current Safety General Requirements due to the used microprocessor technology.

Stage 3.3: Analysis of Fault Detection and Recovery Means

The Analysis of the Fault Detection and Recovery Means is a consequence of the previously mentioned activities, since the implementation of those means can be made by software or hardware technology. The determination of those resources in the system is of fundamental importance, because it influences in the appropriate use of eventual reconfigurations and, in consequence, in the final system modelling.

The influence of fault detection and recovery means in the final modelling of the system can be represented through the concept of Fault Cover Factor. It is not the intention here to determine a numeric value for this Cover Factor, but simply to identify the means of fault/error detection. This determination can be made through the analysis of the involved hardware (FMEA and Improper Inputs) or the analysis of the involved software.

Stage 3.4: Operational Procedures Analysis

Starting from the detailed study of the involved operational procedures, together with the analysis of the hardware and software modules, it takes place an analysis of the operational procedures. This verification tries to identify operational sequences that can lead the system to an unsafe state in the presence or not of faults and/or improper inputs.

Stage 3.5: Interference Analysis of Neighbour Subsystems

Starting from the interface determination of the safety responsible Subsystems, it is performed the determination of certain aspects of the neighbour subsystems that can interfere in an inappropriate way in the safety subsystems. The analysis on these neighbour subsystems should be fundamentally focused on the interference aspects over the vital information of the system.

Stage 4: Safety Final Evaluation

After the accomplishment of the previous stages, it takes place a Final Evaluation of the Safety, standing out:

a) **Hardware**: The conclusions of the Failure Mode and Effect Analysis and of the Improper Inputs for the fail safe and redundant modules are presented, highlighting the faults or improper inputs that can lead the system to an unsafe situation.

b) **Software**: The results obtained in the analysis are presented, showing the problems found according to the following classification:

- Routines that contain problems but that don't put in danger the safety nor the availability of the system;
- Routines that contain problems that can lead the system to anomalies of operation and of availability;
- Routines that contain problems that can lead the system to an unsafe state.

c) **System**: The Degree of Safety of the system is evaluated through the value of its MTTUF - Mean Time To Unsafe Failure, considering also the Fault Cover Factor of the system that is being evaluated. The influence of the operational procedures in the system safety is presented also here.

In this stage, there is a great interaction between the teams of hardware and software analysis. It is calculated also the Reliability of the System, based on the failure rates of its components.

The calculation of the safety degree – MTTUF is based on models that represent the implementation of the system being analysed. The most common models for these evaluations are the Combinatorial Models (Serial and Parallel) and the Markov Models.

Techniques of fuzzy logic are being researched, that can aid in a better evaluation of the safety of the critical system. Besides, it is being researched also a group of quality subfactors that represents the safety quality factor. The following quality subfactors can be pointed out: Precision, Traceability, Completeness, Consistency, Fault/Error Tolerance, Execution Efficiency, Simplicity, Auto-description and Modularity.

Based on the acquired experience in many projects of safety analysis of Metro control systems, it can be pointed that the documentation, generated along the work, is used also by the operator as an instrument for the maintenance of the system and in the new professionals' training. That use is a subproduct of the work of a Safety Analysis.

4. Conclusions

With the growing complexity of the Metro control systems, it must be noted that its safety won't be assured in function of an established technique, but fundamentally due to a Safety Global Program that should be present in all phases of the system life cycle.

In this way, besides the activities of Safety Analysis, the companies should motivate the development of a culture that valorizes and improves the aspects linked to the Safety of the system during all its useful life. They should also avoid the indulgence and then, they must implant a wide quality program.

The return of the investments in safety of Metro control systems, including the activities of safety analysis, is intangible. However, these investments can represent the difference between the peaceful or chaotic operation of the system, between the credibility or the uneasiness.

Based on the obtained practical results, it can be also pointed out that the several stages of our Safety Analysis Methodology, turned to be extremely useful and effective, not discarding, however, the need of a continuous evolution and improvement in the sense of reaching levels of higher safety.

Having that picture in view, the Safety Analysis Methodology has been extremely useful not only improving the safety's factor, but also strengthening the safety's culture disseminated among the members of the process, that is to say, the Operator, the Suppliers and the Independent Safety Analysis Group.

References

1. SAEED, A., LEMOS, R., ANDERSON, T., The role of formal methods in the requirements analysis of safety-critical systems: a train set example. In 21st IEEE International Symposium on Fault Tolerant Computing, Montreal, 1991, p.478-485.

2. LEVESON, N. G.. Safeware systems safety and computers. United States, Addison Wesley Publishing Company, 1995.

3. INTERNET Home Page – http://www.metrosp.com.br/

4. REDMILL, F., CHUDLEIGH, M.F., CATMUR, J.R. Principles Underlying a Guideline for Applying HAZOP to Programmable Electronic Systems, Reliability Engineering and System Safety, n. 55, 1997, p.283-293

5. BROOMFIELD, E.J., CHUNG, P.W.H.. Safety Assesment and the Software Requirements Specification, Reliability Engineering and System Safety, n. 55, 1997, p.295-309

6. LAWRENCE, J.D., GALLAGHER J.M. A Proposal for Performing Software Safety Hazard Analysis, Reliability Engineering and System Safety, n. 55, 1997, p.267-282

7. JAFFE, M.S., LEVESON, N.G., HEIMDAHL, M.P.E.M., BONNIE, E. Software Requirements Analyses for Real Time Process Control Systems, IEEE Transactions on Software Engineering, v.17, n. 3, 1991, p.241-258

8. CAMARGO JUNIOR, J. B. Estudo da Segurança em Sistemas de Controle Metro-Ferroviários, São Paulo, 1996, 142p. Doctoral Thesis (in Portuguese) - Escola Politécnica, Universidade de São Paulo.

9. HAREL, D. Statecharts: a visual formalism for complex systems. Science of Computer Programming, v.8, n.3, p.231- 74, Aug. 1987b.

10. ALMEIDA JUNIOR, J. R. STAD - Uma ferramenta para representação e simulação de sistemas através de statecharts adaptativos, São Paulo, 1995, 202p. Doctoral Thesis (in Portuguese) - Escola Politécnica, Universidade de São Paulo.

Safety Integrity Levels: An Industrial Viewpoint

Vivien Hamilton and Ceri Rees
GEC Marconi Research Centre,
Chelmsford, Essex, U.K.

Abstract

Safety integrity levels (SILs) are now an integral part of safety engineering and appear in almost all modern safety standards. This paper begins with various intuitive interpretations of what SILs mean and how they are calculated. It then considers the engineering basis behind the setting of SILs, paying particular attention to the requirements of the standards IEC 61508, DO-178B, Def Stan 00-56 and Mil-Std 882C.

As is discussed in this paper, SILs are usually used to define the process to be applied in developing the product. It should be remembered, however, that it is the relationship between SILs and the properties of the final product which is actually of interest.

The paper concludes that SILs are a useful informal guide for users and developers for whether a particular product is likely to be suitable for a given application. They also provide a rough guide to the costs of developing a product. However, a rigorous assessment using soundly based product evidence is the only appropriate method for certifying or accepting a product into operational use.

1. Introduction

The concept of safety integrity levels (SILs) appears in some form or other in almost all modern safety standards, particularly those concerned with functional safety of complex systems. However, in any discussion amongst professionals who research or develop such systems, it quickly becomes clear that there is no consensus about what SILs actually mean or how they should be used. Furthermore, examination of the standards reveals that each standard has a subtly different philosophy behind assignment and use of SILs. This leads to the conclusion that, without first revising the standards to some common framework, there can never be any simple, generally applicable mapping between the SILs of the different standards.

The word engineering implies the application of principles, founded on appropriate scientific and technological disciplines. When SILs are used in an engineering context, it is essential to ensure that their philosophy has a sound scientific basis and that they are applied in a disciplined manner. The fact that this is not always the case has led to a recent paper [Lindsay 1997] in which the current reliance on SILs in standards is criticised. One of the authors of that paper has also described SILs as a 'self-contained calculus of warm glows', a description which aptly captures the often subjective and circular arguments which rely on them.

In this paper the topic of SILs is approached from a industrial viewpoint. The paper begins with a basic guide to SILs and then explores some of the distinctions between the different definitions. The paper is intended for an audience with a background in both safety engineering in general and the development of safety related software in particular. It therefore assumes that the reader will be familiar with concepts such as *hazard analysis, risk assessment, safety cases, formal methods* and *statistical testing*. Readers without this background are referred to the glossary and to the references given.

1.1 A Rough Guide to SILs

Most people would agree that the greater the safety risk if a system malfunctions, the greater the effort that should be devoted to creating a failure free product. This is the principle behind SILs which are used to define what measures should be taken in the development of a product to reduce the risk that failures arising from design defects or other systematic faults will create a safety hazard.

The most widely used SIL scheme has five broad bands, where the bottom band, SIL0, applies to systems that are not safety related and the top band, SIL4, applies to safety critical systems. Since the standards are intended to apply to safety related systems, SIL0 is largely ignored. In IEC 61508 [IEC 1998], and in this paper, the bands are termed SIL1, SIL2, SIL3 and SIL4. Slight variations on this terminology are used elsewhere, e.g. Def Stan 00-56 [MOD 1996] uses S1 to S4. A rough interpretation of the meaning of each SIL is provided in figure 1.

In the rough guide to SILs, each SIL is characterised in two different ways: by how reliable the system needs to be and by how much effort should go into developing the system. Standards from diverse industry sectors place different emphasis on each of these factors and have different expectations of how much effort (and what methods) should be used to achieve high integrity levels. These differences will be explored further in section 2 of this paper.

A Rough Guide to SILs	
SIL0	System does not affect safety. No assumptions are made about its dependability. No safety requirements are imposed on how it is developed.
SIL1	System developed to normal commercial grade of quality and technology. Such systems are expected to fail about once in 10 demands or about once in 10 hours continuous use. SIL1 systems are not expected to be the direct cause of a fatal accident, but they could be a minor contributing factor, e.g. by causing a nuisance to operators of safety critical equipment.
SIL2	High quality commercial (or bespoke) systems. Such systems are expected to fail about once in 100 demands or about once in 100 hours continuous use. With careful system design, it should be possible for very good quality commercial off the shelf (COTS) products to be justified for use in SIL2 systems. SIL2 systems are not expected to be the sole cause of a fatal accident but they could provide a significant contribution, e.g. by providing erroneous information to operators of safety critical equipment.
SIL3	System specifically developed to meet safety requirements. The development should use specialist methods and techniques (e.g. for software, formal methods and safe subsets of languages should be used) and the highest levels of QA, modelling, analysis and testing. SIL3 systems may be dependable to a level of one failure for every 1,000 demands (or 1,000 hours continuous use). It is most unlikely to be feasible to use COTS products (except for off-the-shelf products specifically designed for safety applications) for the implementation of SIL3 functions. Although failure of a SIL3 system should not be the sole cause of a fatal accident, it may create a situation which challenges the safety protection systems.
SIL4	The very highest grade of system of proven quality utilising multiple hardware redundancy. Software able to meet this level of integrity demands the highest possible project competence, the best available technology, the closest attention to the correctness of the requirements, design and code, and analysis of the code to demonstrate it matches the requirements, in addition to all other elements of good practice. SIL4 systems may be dependable to a level of one failure for every 10,000 demands (or 10,000 hours continuous use). SIL4 systems are safety critical.

Figure 1 - A Rough Guide to SILs

2. Setting the SIL Requirement

Traditionally, the reliability of a system is estimated by compounding the experienced fault rates of the components to arrive at the overall random failure rate. This traditional approach takes no account of failures due to design faults. For systems containing software (or indeed other types of complex systems) the majority of failures, although dependent on external circumstances, are not caused by random factors or wear-out but by systematic factors arising from human error in the development or operational life of the equipment. For such systems, the traditional approach to reliability cannot be used.

The control of systematic factors affecting reliability, whether in hardware or in software, depends on understanding the ways in which faults can arise in the delivered system. Such faults basically arise from human errors which have been introduced during the development process and have remained undetected until the system is used. These faults may originate in the statement of requirements, or in the design and implementation of the system. In order for a failure to occur in use, the system must encounter circumstances which challenge the faults. Even when faults are present and circumstances occur which challenge the faults, a failure (or at least a dangerous failure) may not result if the design is fault-tolerant, for example with changeover capacity or with redundant and diverse backing systems.

SIL requirements are typically determined from an application-dependent assessment of the tolerable risk of the hazards associated with the application. The tolerable risk is that risk generally accepted by the public, or by a regulating agency on their behalf, for the consequences of the hazards. The tolerable risk can then be translated into a SIL requirement using the tables in standards such as Def Stan 00-56 or IEC 61508.

2.1 IEC 61508

IEC 61508 requires the SIL for a particular system to be determined from hazard analysis which explores the system and its interaction with its environment. In IEC 61508 it is assumed that the system comprises some equipment which is being controlled and which presents a hazard (e.g. a chemical reaction which might overheat or a piece of machinery which might move in a dangerous manner). Hazard analysis should determine the likely consequences if the equipment goes out of control and the probability of it doing so. The combination of consequence and probability (called the risk) should be assessed. This may indicate that the control of the equipment presents an intolerable risk and that therefore a protection system is needed to reduce it. The protection system will normally act to reduce the probability of the equipment going out of control (e.g. by shutting it down). The factor by which this probability needs to be reduced to make the risk tolerable therefore determines the reliability required of the protection system. If the protection system is a simple mechanical (or electrical) device, the reliability requirement will relate to traditional equipment reliability based on random failure and wear-out mechanisms. If however, the protection system is a complex system (e.g. a software system), the reliability requirement will need to be translated into a SIL. The SIL requirement is best determined from a quantitative risk assessment process. If however numeric data is not available a qualitative (pseudo-quantitative) process can be used in which words such as *frequent* and *occasional* are used to represent order of magnitude bands of probability. Tables in IEC 61508 provide guidance on how to relate SIL to reliability.

IEC 61508 is intended as a generic standard from which sector specific standards can be developed. If there is no specific standard for a particular industry, IEC 61508 can also be used by individual projects as a standard in its own right. Unfortunately, it does not follow that all industries using IEC 61508 (or instantiations of it) will have a common interpretation of SILs. The tables which relate reliability to SIL are provided for guidance rather than being normative: it is possible for two different industries to equate different reliabilities to the SILs. Furthermore, it is also possible within the requirements of IEC 61508 for a different SIL scheme to be applied (such as the one defined in Def Stan 00-56 which is described below).

2.2 DO-178B

DO-178B [RTCA 1992] is actually a guidance document defining recommended practices for developing software for the avionics industry. In practice, for avionics software, it is difficult to ignore its recommendations and hence DO-178B is normally treated as a standard. The assignment of software levels (the equivalent of SILs) in DO-178B is based upon first assigning failure conditions to system functions. There are five failure condition levels ranging from *Catastrophic* (failure conditions which prevent continued safe flight and landing) through *Minor* (failure conditions which would not significantly reduce aircraft safety and which would involve crew actions which are well within their capabilities) to *No Effect*. DO-178B also defines five software levels ranging from Level A software whose anomalous behaviour would cause or contribute to a catastrophic failure condition to Level E which applies to software whose failure would have no effect on aircraft capability or pilot workload. Software systems are assigned to these five levels according to the consequence of the software failure i.e. level A software is flight critical software whose failure would be likely to result in loss of the aircraft. Level E software is not safety related since its failure would not affect the safety of the aircraft.

At first glance the consequence based software levels in DO-178B and the reliability based assignment of SILs in IEC 61508 appear to be assigned on orthogonal scales. There is, however, a relationship between reliability and consequence which is implicit in the hazard analysis undertaken for avionics systems. A reliability requirement has to be assigned to avionics systems which enables the overall aircraft safety targets to be met. In the system level guidance document related to DO-178B, ARP 4754 [ARP 1996], reliability requirements are mapped to failure condition levels. Systematic failures should be managed by an appropriate design and assurance approach so that the actual failure rate due to systematic failures is not significant compared with random failures. For this to be the case, it would be reasonable to assume that systematic failures should account for no more than ten per cent of the total failure rate. In ARP 4754, each failure condition level spans two orders of magnitude and level A under this scheme is expected to be five orders of magnitude more reliable than SIL4 in IEC 61508.

2.3 Def Stan 00-56

Def Stan 00-56 assigns SILs using a scheme which considers both consequence and reliability. In Def Stan 00-56, five SILs are used, but these are abbreviated to S0 to S4. The highest SIL, S4, is required for functions which provide the primary safeguard against consequences resulting in one or more deaths. If the primary function to prevent this type of accident has less than the required reliability then a S4 backup function will be required. Primary functions with greater reliability can have backup functions of lower SILs. The SIL required of the backup function is therefore dependent not just on the possible consequences, but also on how often the backup is likely to be relied upon. The lower SILs (S1, S2 and S3) also apply to primary functions where the consequence is less than death.

In Def Stan 00-56 there are therefore two different schemes for assigning SILs depending upon whether the function is the primary safety function or is the second (or subsequent) safety function.

Def Stan 00-56 also allows a function to be implemented by two or more independent lower SIL functions, provided that the combining element is of the required integrity. For example, a S4 function could be implemented by two independent S3 functions with a S4 combining element, such as a voting circuit.

The concept of claim limits based on SILs also exists in Def Stan 00-56. Claim limits need not be used when there is actual data available from real experience of operating the system. Otherwise, claim limits, as the term suggests, limit the claim for reliability that can be made for a piece of equipment, based on its SIL. Hence the claim limits map reliabilities to SILs.

2.4 Mil-Std 882C

The philosophy in Mil-Std-882C [DOD 1993] is that traditional methods for estimating reliability used for other types of technology are not appropriate for software systems. Mil-Std 882C uses the concepts of a sofware control category and software hazard risk index. Four levels of software control category are used depending on whether failure of the software could directly lead to the hazard, whether there is the potential for other systems (or people) to provide mitigation or whether there are several, redundant, independent safety measures for each hazardous event. The software control category is then combined with consequence in a table in a similar manner to the way in which Def Stan 00-56 combines probability with consequence. The combination of software control category and consequence is used to define a software hazard risk index (ranging from 1, high risk to 5, low risk). Although Mil-Std 882C does not give details, except at a high level of abstraction, the process requirements for these software hazard risk indices are similar to the process requirements imposed by DO-178B and Def Stan 00-55.

2.5 Summary of philosophies in standards

Using these four standards as examples, it can be seen that there are different philosophies for the assignment of SILs. IEC 61508 maps SILs to reliability; DO-178B maps SILs to consequence and therefore to the confidence that is needed in the system's ability to fulfil its safety functions and Def Stan 00-56 utilises a hybrid of these views. Within Def Stan 00-56, tables for secondary functions and for claim limits imply that SILs can be mapped to reliability whereas the table for primary functions requires a minimum assurance level based on consequence alone.

Outside the defence and avionics industries, other standards also adopt similar philosophies to one or more of the standards discussed above. For example, in the motor car industry, the MISRA guidelines [MISRA 1994] define levels of controllability in a similar manner to the software levels of DO-178B. In the MISRA scheme, the levels map only to consequence, i.e. a failure of the highest controllability level software could lead to a situation in which the occupants (of the vehicle) would be unable to control the vehicle.

Because different standards have different philosophies underpinning the SIL schemes and different members of the safety community have different expectations of what SIL means, there is no consensus on mapping between consequence, probability and SIL. In addition, there is no general mapping between the levels of the different SIL schemes.

3. SILs and the Process

Having determined the SIL for a system, most standards provide normative requirements or guidance for the techniques which should be applied to the development of the system. In general, as the SIL increases, the system developer is expected to do more to avoid dangerous failures and to carry out the normal development activities more thoroughly.

For safety related systems most standards recommend that there is some form of defence in depth in the final product. Redundant hardware can be used to protect against random hardware failures and functional diversity can be used to protect against the consequences of systematic faults. For example, as previously discussed, Def Stan 00-56 permits a function of one SIL to be implemented by two or more independent lower SIL systems. A well implemented defence in depth strategy will generally be more effective in both risk and financial terms than relying on a single, very high integrity function. Whatever the approach, development of high integrity systems still requires more work (e.g. to develop the additional, diverse or redundant systems) compared with development of low integrity systems.

The goals of a safety related system development lifecycle may be summarised as follows:

1. Define safety requirements and gain confidence that the safety requirements are necessary and sufficient for safety.

2. Achieve the development of a system which fully satisfies the safety requirements under all conditions which the system will encounter (including reasonably foreseeable misuse).

3. Ensure that any functions or behaviours that the system is required to have in addition to the safety requirements do not compromise the satisfaction of the safety requirements.

4. Provide evidence that goals 1 to 3 have been achieved.

The development activities for safety related systems (especially for high SIL systems) are also typically required to be more rigorous than those used in the development of other systems. For example, Def Stan 00-55 [MOD 1997] requires the safety requirements to be specified using formal methods. Requirements for configuration management and quality assurance activities are also likely to be more demanding.

A defence in depth strategy can be applied to the verification and validation activities which take place within the development lifecycle. Within the development lifecycle, fault detection can be maximised by selection of complementary techniques which target different classes of error and by using independent persons or teams to protect against common blindspots. For examples of this type of approach see Def Stan 00-55 in which both static analysis techniques and dynamic testing techniques are required. Def Stan 00-55 also requires a V&V team which is independent of the design team.

The development of high SIL systems requires more effort than the development of low SIL systems and hence is more expensive. Therefore SILs can be mapped to the cost of developing the system. However, even this relationship is complicated by the fact that some of the additional effort may be cost effective in that it results in the elimination of errors at an early stage in the lifecycle. The cost effectiveness of 'left shift' activities is a common theme in software development. For example, organisations that have adopted the Capability Maturity Model[1] [SEI 1995] (CMM) have provided evidence that they have gained a considerable return on investment [Watts 1991]. This is despite the fact that the CMM is fairly demanding in its expectations of requirements management, configuration management and software quality assurance even at its first rated maturity level.

Standards (e.g. IEC 61508) generally relate activities or methods to SIL by means of tables where the particular methods are rated HR (highly recommended), R (recommended), NR (not recommended) and sometimes M (mandated). The difficulty with this type of classification scheme is that some methods will be complementary while others will overlap. If the standard does not give requirements on how to select the overall package of methods, then compliance to the standard overall is ambiguous.

Def Stan 00-55 adopts a different approach, using M (mandatory), J1 (justification is required if the 'letter' of the standard is not followed) and J2 (justification is required if the 'spirit' of the standard is not followed). This justification approach still requires judgement in determining whether the justification is appropriate and adequate, but at least the justification is required to be recorded and is therefore able to be scrutinised. In any safety related development, it is the unrecorded assumptions and justifications which present the greatest risk of undermining the safety case.

Relating SILs to process requirements sometimes leads people to put forward a spurious safety argument as follows:

- I need a SIL4 system; therefore I'll follow the standard for a process to develop a SIL4 system.
- I've followed the standard for a process to develop a SIL4 system; therefore I now have a SIL4 system.

This is not an adequate safety argument. The misunderstanding that causes such arguments to be put forward appears to arise from confusion between the twin goals of any safety engineering activity. The aims are to engineer a safe system and to provide evidence that it is as safe as it ought

---

[1] Capability Maturity Model and CMM are trademarks of the Software Engineering Institute at Carnegie Mellon University.

to be. The role of standards, SILs and process in producing a safe system can be illustrated as follows:

- I need a SIL4 system; good practice dictates that I follow the standard for a process to develop a SIL4 system.
- In following the standard I've been forced to generate the evidence necessary to demonstrate that the system is adequately safe; assessment of the evidence shows that I have a SIL4 product.

Standards generally provide a process which provides guidance and sometimes constraints that should help in developing a high SIL system. The perfect standard would define what evidence is needed but leave the developer free to decide how to engineer the product and generate the evidence. Unfortunately a standard which does not constrain the development process in any way is probably impossible to write. It would be extremely difficult to define unambiguously the evidence which is both necessary and sufficient for all cases. Also, it is unlikely that such a standard would be well received because most users would find a standard that lacked guidance on the process difficult to apply.

4. SILs and the System

While the standards generally emphasise *process*, it is of course the system or final *product*, which is really of interest. It is the system which may fail with dangerous consequences. From a commercial point of view, the customer pays for the qualities of the system even if it is the special qualities of the process which determine the cost.

Hence safety cases (or other forms of certification or assurance evidence) should generally use arguments based on observed or analysed properties (testing, modelling or analysis) of the system.

The best indicator of reliability for any system is hard statistical evidence of its actual performance. A prediction of the system's expected performance can be derived from statistical modelling using test results (e.g. from failure free random testing, or from reliability growth modelling of mean time between failures). Such statistical testing should be backed up by other evidence to provide confidence in the validity of the assumptions (and statistical model). Deterministic evidence (e.g. proof of absence of statically determinable faults) is also desirable to provide confidence that there are no other failure modes which have not been considered.

The value of measurement is captured in the following quote by Lord Kelvin:

"When you can measure what you are speaking about and express it in numbers you know something about it; but when you cannot measure it, when you cannot express it in numbers, your knowledge is of a meagre and unsatisfactory kind."

Statistically valid methods, such as random testing are generally intended as assurance activities. Because they are intended for assurance purposes, it is often intended that no failures should be encountered during the random testing. This may be essential if a claim for very high reliability is being made. However, if faults do exist in the system, it is better for them to be challenged during random testing than for the system to fail in use.

Random testing sometimes also has a value in 'debugging' a software system. Even when large amounts of verification and validation effort have already been expended within the development process, faults may still be detected during random testing. One reason for this is that the random testing should be free of the implicit assumptions which may have affected hand crafted tests, manual inspections and analyses conducted earlier in the development process. So although the developers should aim to have failure free random testing, detecting faults at this stage is better than not detecting them. If faults are found during assurance activities then great care should be exercised in determining whether these have arisen because of an erroneous implicit assumption.

There are problems in assessing (or providing) evidence that a system meets a high integrity requirement particularly when this relates to a high reliability requirement. It is generally not practical to demonstrate (by reliability growth modelling or statistical testing) that a system achieves a reliability of greater than around about 10^{-4} (failures per demand or failures per hour). To demonstrate such figures statistically would typically require many years worth of testing.

Paradoxically, standards generally do not mandate statistical methods for low SIL (e.g. SIL1 or SIL2) systems, where a statistical determination of reliability may be feasible. Such methods do however tend to be recommended for high integrity (SIL3 and SIL4) systems, where it may be infeasible to provide sufficient testing to demonstrate the required reliability.

Deterministic evidence can in principle be used to construct a causal model of how a system can and can not fail and hence provide a strong safety argument that the system can not fail in a dangerous manner. However, generation of deterministic evidence for software systems generally requires that the software is designed within certain style and structure constraints which make analysis feasible. The lack of such constraints may mean that use of deterministic evidence is impracticable for the justification of many software systems. For SIL3 and SIL4 systems, the standards often impose these constraints on the development process.

Standards writers and system developers typically seek to constrain the process in the hope that by doing so they can influence the properties of the final system. Unfortunately it is generally difficult (or impossible) to provide cause and effect links between process deviations and specific faults in the final system. This means that it is not possible to use process to predict the failure rate of the final system.

The fact that knowledge of the process provides very little evidence of the actual properties of the final system can be illustrated by considering that the actual properties demanded of high integrity systems can vary widely. For example, it is possible to have a system with a very high availability requirement for which the occasional individual erroneous output can be tolerated. At the opposite extreme it is also possible to have a system for which there are no dangerous consequences from unavailability but for which all outputs, whenever the system is operating, have very high reliability requirements. Systems of each type could be developed using identical processes, it is only in the detailed application of the development activities where there would be differences.

Sometimes a developer may seek to justify that a system is high integrity on the basis that it has been developed using a process which has a track record of being used for developing high integrity systems. Such claims usually provide virtually no evidence for system integrity and at worse may be based on a fallacy. For example, suppose a supplier has developed a high integrity system before and intends to use the same process to develop another high integrity system. The track record of the single successful example does not provide adequate statistical evidence to make a meaningful claim about the probability of success.

Claims that following some standard will result in a very high integrity system should always be viewed with suspicion. Such claims are particularly misleading when the supplier concerned is from outside of the industry which has a successful track record of using the standard (when use of DO-178B is quoted outside the avionics industry for example). Standards are only part of a much wider industry framework, incorporating, for example, engineering culture, regulatory processes and application specific precedents.

5. Using SILs in Practice

In general, while the concept of SILs provides a useful engineering shorthand for the level of dependability to be expected from a system, SILs do not map well to any particular properties of particular systems. Quoting SILs is a little like a car salesman quoting insurance classes. Insurance class is a weak indicator of the features of the car but provides an idea of how much the insurance will cost. In much the same way, SIL provides an indication of the level of effort that has gone into removing design errors, but is a poor predictor of system reliability.

There appears to be no sound method for linking SIL with the actual properties of the final system, such as reliability. There is justification in linking SILs to consequence or to risk (the combination of consequence and probability) since SILs then provide a means of prioritising risk reduction actions. This prioritisation can then be used when reducing risk to a tolerable level and then further reducing risk until it is as low as reasonably practicable (ALARP).

The link between process activities and SILs can be used to provide an industry consensus on best practice (this is the case with DO-178B) or a definition of the customer's expectation of what suppliers should be required to do (as in the case of Def Stan 00-55). Such a consensus on expected practice is consistent with the concept of 'reasonably practicable' within the ALARP approach.

It is not obvious however, that there can then be a consistent mapping of SILs across different industries, which tend to have different expectations of what is reasonable and what is achievable.

The defence industry has equipment requirements that span such diverse applications as road transport, air transport and nuclear power plant. In the defence industry therefore, the lack of consistency in SILs is likely to mean that either SILs are inconsistently used across defence applications or that SILs are inconsistent between defence and equivalent civilian applications.

Linking SIL with consequence occasionally leads to the situation where a system is required to have a high SIL, but a modest reliability target. In such cases, the rationale for the high SIL is not to achieve high reliability but to have additional confidence in the validity of the stated reliability. For a high SIL system there will need to be confidence that the assumptions made in measuring the reliability are valid and that the residual doubt arising from the potential for unconsidered failure modes has been minimised and bounded.

From the engineering point of view the consequence linked scheme of DO-178B (or the controllability type scheme of Mil-Std 882C) has the advantage of being easy to apply and intuitive for the domain experts developing the system. An intuitive scheme helps to focus attention on the aspects of the problem that really matter; the potential safety issues.

Since the process and the SIL are weak indicators of system properties, quoting the SIL does not in itself provide a good safety argument. Process type SIL arguments should not be used as a primary safety argument, instead emphasis should be placed on demonstration of the adequacy of safety in the final system. A SIL argument, however, may form a useful argument for the reduction of residual doubt.

Where SILs are linked to process, it is important that measurement is used to close the loop. For example, Def Stan 00-55 requires that developers define the process they intend to use and also the metrics that will demonstrate that the process is performing as expected.

For the customer considering purchasing a high integrity system (or a supplier intending to reuse a previous product in a new application), SIL can only be used as an informal classification system. An informal scheme has some value, for example in providing a quick check of feasibility. A detailed comparison of the requirements of the application and of the actual properties of the proposed system may however reveal that the 'integrity' is incompatible.

6. Conclusions

Although SILs appear in almost all modern safety standards, there is no consensus about what they actually mean or how they should be used.

Linking SILs with consequence and with industry consensus on what is reasonably practicable appears to be more soundly based than attempting to link SILs with reliability. However, SILs can support a reliability claim based on other evidence by adding confidence about the level of residual doubt in the reliability claim.

SILs do not provide a sound basis for safety arguments and therefore evidence, rather than SIL, should be used to support the claim for the properties of the final system.

SILs do provide a useful informal categorisation scheme for purchasers and developers of safety related systems. As an informal scheme they can be useful for the following:

- As a rough guide for whether an existing system is likely to be suitable for a particular safety related application.
- As a guide to industry consensus on what is reasonably practicable.
- As a rough guide to the likely costs of developing a safety related system.

Glossary

ALARP	As Low As Reasonably Practicable. A philosophy applied to the reduction and acceptability of risk. An explanation is provided in [HSE 1992].
formal methods	Methods for the specification and production of software based on a mathematical system. Formal methods permit mathematical reasoning about the specification.
functional diversity	The same high level requirement achieved by two or more different and independent functions implemented in separate and independent systems.
hazard analysis	Systematic determination of all possible accidents, consequences and initiating events along with their probabilities.
risk assessment	Assessment of the system or a component to establish whether the achieved risk is lower than or equal to the tolerable risk level.

safety cases	Collective term for the body of evidence including reports which provide a reasoned justification, supported by evidence that the system is tolerably safe. A discussion on the structure of claims in a safety case is provided in [SCSC 1998].
static analysis	The activity of evaluating software without actually executing it. Static analysis can include simple activities such as the activity of the compiler in checking that the code uses the correct syntax. In high integrity software it generally refers to highly specialised techniques. An explanation of the techniques and their benefits is provided in [SCSC 1998].
statistical testing	Testing conducted with the intent of performing a statistical analysis of some property of the system, usually reliability. Such testing is generally very extensive, includes randomisation of test values and requires specialised automated test equipment.

References

[ARP 1996]. ARP4754: *Certification Considerations for Highly-Integrated or Complex Aircraft Systems.* Aerospace Council, SAE, 1996.

[DOD 1993] Military Standard 882C: *System Safety Program Requirements.* Department of Defense, 1993.

[HSE 1992] Health and Safety Executive: *Tolerability of Risk in Nuclear Power Stations.* HMSO Revised 1992.

[IEC 1998] IEC 61508 Committee Draft for Voting 1998: *Functional Safety of Electrical/Electronic/Programmable Electronic Safety Related Systems.* IEC 1998.

[Lindsay 1997] Lindsay P A & McDermid J A: *A Systematic Approach to Software Safety Integrity Levels in SafeComp '97 Proceedings of the 16th International Conference on Computer Safety, Reliability and Security, York, 7-10 September 1997, edited by Peter Daniel.* Springer-Verlag London Limited, 1997.

[MISRA 1994] The Motor Industry Software Reliability Association (MISRA): *Development Guidelines for Vehicle Based Software.* Motor Industry Research Association, 1994.

[MOD 1996] Defence Standard 00-56/2: *Safety Management Requirements for Defence Systems, Issue 2.* Ministry of Defence, 1996.

[MOD 1997] Defence Standard 00-55/2: *Requirements for Safety Related Software in Defence Equipment.* Ministry of Defence, 1997.

[RTCA 1992] RTCA/DO-178B: *Software Considerations in Airborne Systems and Equipment Certification*. RTCA 1992.

[SCSC 1998] Safety-Critical Systems Club (SCSC): *Industrial Perspectives of Safety-Critical Systems, Proceedings of the sixth Symposium Birmingham 1998*. Springer-Verlag, 1998.

[SEI 1995] Software Engineering Institute, Carnegie Mellon University: *The Capability Maturity Model: Guidelines For Improving The Software Process*. Addison Wesley, 1995.

[Watts 1991] Watts H S, Snyder, T R & Willis, R R: *Software Process Improvement at Hughes Aircraft in IEEE Software, vol. 8, no. 4, July 1991 pp 11-23*. IEEE, 1991.

Acknowledgements

Industrial experience arises from a variety of sources including direct involvement in building and assessing systems and discussion with colleagues and fellow professionals. The authors wish particularly to acknowledge private discussions with Professor John McDermid of the University of York, with Dr. Keith Rodger of DERA Boscombe Down, with Professor Bev Littlewood of City University and with Professor Robin Bloomfield and Dr. Peter Froome of Adelard. Discussions with fellow authors of Def Stan 00-55/2 and members of the Advisory Committee on Safety of Nuclear Installations working group on the Safety of Operational Computer Systems also formed a part in shaping the concepts in this paper. The authors are also grateful to all the contributors to the discussions on SILs which have taken place on the York high integrity systems e-mail discussion group, moderated by Dr. Jonathon Moffat and on the IEEE software standards e-mail working group led by Dr. Victoria Stavridou.

Code generation in the SACRES project *

Thierry Gautier and Paul Le Guernic **

IRISA / INRIA, Campus de Beaulieu – 35042 Rennes Cedex – FRANCE
e-mail: {Thierry.Gautier|Paul.LeGuernic}@irisa.fr
http://www.irisa.fr/EXTERNE/projet/ep-atr

Abstract. The SACRES project is dealing with the development of new design methodologies and associated tools for safety critical embedded systems. Emphasis is put on formal techniques for modular verification of the specifications, distributed code generation, and generated code validation against specifications. This is allowed by using a single formal model which is that of the DC+ format, which provides a common semantic framework for all tools as well as end user specification formalisms. Modular and distributed code generation is the main subject of this paper. Distributed code generation aims at reducing the dependency of the design with respect to the target architecture. Modularity helps reuse of existing designs, and makes it possible to address much larger systems.

1 Introduction

The overall objective of the SACRES project is to provide designers of embedded control systems, in particular safety critical systems, with an enhanced design methodology supported by a toolset significantly reducing the risk for design errors and shortening overall design times. This is achieved through the use of the maximum degree of automation, especially in respect of code generation and verification [6]. SACRES architecture is depicted in figure 1. The SACRES toolset combines the following main groups of tools:

- The command-level interface permits the tools to be invoked. Some of the tool launch facilities are also available through the menus from the specification tools.
- The specification front-end tools are mostly self-contained graphical tools for building models. As software requirements for embedded systems typically involve a mixture of state oriented and dataflow oriented descriptions, SACRES offers the possibility of *mixed formalism* design, by combining expressive power of Statecharts [12] from the STATEMATE tool, and SIGNAL

* This work is supported by the Esprit project R&D SACRES (EP 20897). SACRES members are: Siemens (Lead partner), i-Logix, TNI (Techniques Nouvelles d'Informatique), OFFIS, INRIA, the Weizmann Institute of Science, British Aerospace, SNECMA.

** The following people have also participated to this work: Albert Benveniste, Loïc Besnard, Patricia Bournai, Sylvain Machard, Éric Rutten. They are gratefully acknowledged.

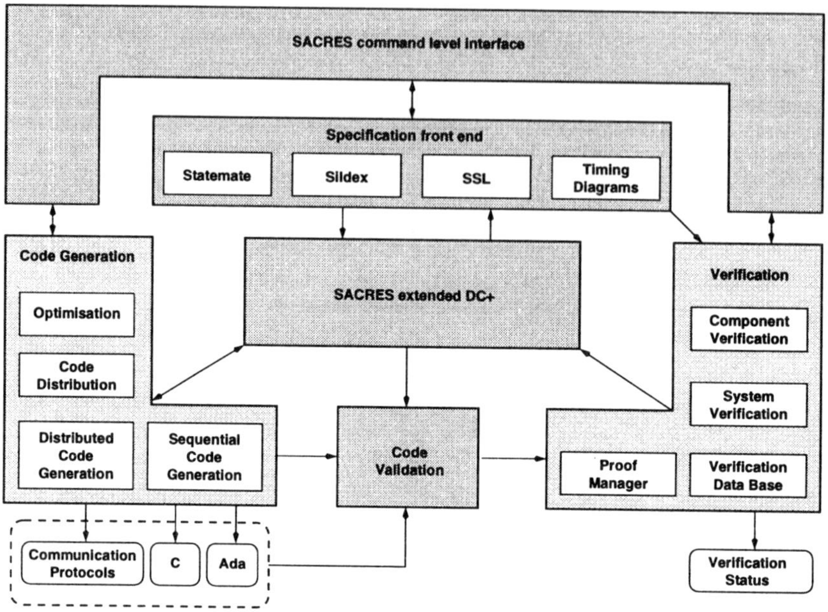

Fig. 1. SACRES *architecture*

[15] based dataflow diagrams from the SILDEX tool. SSL is a textual language, used for assembling models with components from different specification tools. It is mainly available from the graphical editors of the specification tools.

- The DC_+ representation is a global model format [19]. These files are not expected to be read by the user, just to be passed between tools in the toolset.
- Code generation tools include an interface for defining target architecture and code distribution as well as the code generation engine itself.
- Code validation provides a formal correctness check for the code generation translation.
- The verification tools allow automatic checking of the models that are produced from the specification tools, and manage the results of the proof.

This paper is dedicated to a presentation of the methodology adopted in SACRES for *modular and distributed code generation.* The main add-on of the methodology is to allow *automatic generation* of efficient distributed code from the specification, entirely replacing the manual coding phase still employed in current industrial design flows. A requirement for that is, at the front-end level, the use of specification tools that are based on a formal semantical model. This is the case of both Statecharts and SIGNAL formalisms. Further, since *mixed designs* are favoured in SACRES, using together both state-based and dataflow specification styles, a common representation was in some way mandatory. In

SACRES, this common representation is the $\text{D}_{\text{C}+}$ format. This format implements the paradigm of *synchronous programming* [8,11] in its full generality. Although very close in its syntax of the *synchronized dataflow model* advocated by the SIGNAL language, it constitutes a model for the semantic integration of SIGNAL [10] and Statecharts specifications [13]. In particular, translations from Statecharts to $\text{D}_{\text{C}+}$ have been defined in SACRES [4,5].

The semantical basis of the $\text{D}_{\text{C}+}$ format is that of *Symbolic Transition Systems* presented in the next section. This model includes in particular scheduling specifications, which are used to represent causality relations, schedulings, and communications. The compilation of programs expressed in $\text{D}_{\text{C}+}$ results in a hybrid structure which allows to represent both control and scheduling of the programs, and to study important properties such as *endochrony.* Part of this structure is a directed graph with boolean conditions on nodes and dependences. From section 3, we will use a more macroscopic view of this directed graph, with the help of diagrams, to present the methodology of code distribution. In section 3, we give an overview of the approach and in section 4, we present the main steps of the method.

2 The DC+ model

SACRES relies on a strong formal modelling basis, which is that of Symbolic Transition Systems with Scheduling specifications [17,18]. This abstract model is used as the model of the $\text{D}_{\text{C}+}$ concrete format.

2.1 Symbolic Transition Systems with Scheduling specifications

Symbolic Transition Systems We assume a vocabulary $\mathcal{V}$ which is a set of typed variables. All types are implicitly extended with a special element $\perp$ to be interpreted as "absent". Some of the types we consider are the type of *pure flows* with domain $\{\text{T}\}$, and *booleans* with domain $\{\text{T},\text{F}\}$ (recall both types are extended with the distinguished element $\perp$).

We define a *state* s to be a type-consistent interpretation of $\mathcal{V}$, assigning to the set of all variables, a value for it over its domain. For a subset of variables $V \subseteq \mathcal{V}$, we define a V-state to be a type-consistent interpretation of V. Thus a V-state assigns to the set V a value $s[V]$ for it over its domain; also, for $v \in V$ a variable, we denote by $s[v]$ its interpretation by state s.

We define a *Symbolic Transition System* (STS) to be a system

$$\Phi = \langle V, \Theta, \rho \rangle$$

consisting of the following components:

- V is a finite set of typed *variables,*
- $\Theta(V)$ is an assertion characterizing *initial states.*

- $\rho = \rho(V^-, V)$ is the *transition relation* relating past and current states s^- and s, by referring to both past and current versions of variables V^- and V. For example the assertion $x = x^- + 1$ states that the value of x in s is greater by 1 than its value in s^-. If $\rho(s^-[V], s[V]) = \text{T}$, we say that state s^- is a *ρ-predecessor* of state s.

A *run* $\sigma : s_0, s_1, s_2, \ldots$ is a sequence of states such that

$$s_0 \models \Theta \ \bigwedge \ \forall i > 0 \ , \ (s_{i-1}, s_i) \models \rho$$

The *composition* of two STS $\Phi = \Phi_1 \bigwedge \Phi_2$ is defined as follows:

$$V = V_1 \cup V_2 \ , \Theta = \Theta_1 \wedge \Theta_2 \ , \rho = \rho_1 \wedge \rho_2 \ ,$$

the composition is thus the pairwise conjunction of initial and transition relations. Note that, in STS composition, interaction occurs through common variables only. Hence variables that are declared private to an STS will not directly contribute to any external interaction.

Notations for STS*:*

- $c, v, w, \ldots$ denote STS *variables,* these are the declared variables of the STS; useful additional variables are the following:
- for v a variable, $h_v \in \{\text{T}, \perp\}$ denotes its *clock:*

$$[h_v \neq \perp] \Leftrightarrow [v \neq \perp]$$

- for v a variable, ξ_v denotes its associated *state-variable,* defined by:

$$\textbf{if } h_v \textbf{ then } \xi_v = v \ \textbf{ else } \xi_v = \xi_v^-$$

  Values can be given to $s_0[\xi_v^-]$ as part of the initial condition. Then, ξ_v is always present after the 1st occurrence of v. By convention, ξ_v is private to the STS in which it is used. Thus state-variables play no role in STS-composition. Also, note that $\xi_{\xi_v} = \xi_v$, thus "state-state-variables" need not to be considered.

Transition relations for STS are naturally specified using conjunction of predicates.

Modularity. As modularity is wanted, it is desirable that the pace of an STS is local to it rather than global. Since any STS is subject to further composition in some yet unknown environment, this makes the requirement of having a global pace quite inconvenient. This is why *we prohibit the use of clocks that are always present.* This has several consequences. First, it is not possible to consider the "complement of a clock" or the "negation of a clock": this would require referring to the always present clock. Thus, clocks will always be variables, and we shall be able to relate clocks only using $\wedge$ (intersection of instants of presence), $\vee$ (union of instants of presence), and $\setminus$ (set difference of instants of presence).

Scheduling specifications Modular code distribution, and in the same way, separate compilation, clearly require to be able to reason about causality, schedulings, and communications. This is why we enrich the STS model as follows.

Preorders to model causality relations, schedulings, and communications. We consider again a set V of variables. A *preorder* on the set V is a relation (generically denoted by $\preceq$) which is reflexive ($x \preceq x$) and transitive ($x \preceq y$ and $y \preceq z$ imply $x \preceq z$). Preorders are naturally specified via *(possibly cyclic)* directed graphs:

$$x \rightarrow y \quad \text{for} \quad x, y \in V \ .$$

The *conjunction* of two preorders is the minimal preorder which is an extension of the two considered conjuncts.

A *labelled preorder* on V is a preorder, together with a domain for each $v \in V$. Call $\mathbf{dom}_V$ the domain of the set V of variables. Denote by Λ_V the set of all labelled preorders on V. A *state* s is now a type consistent interpretation of the labelled preorder, i.e., a preorder on V together with a value $s[V]$ for the set of all variables belonging to V. Denote by $\mathbf{dom}_{\Lambda_V}$ the domain in which states take their value.

STS *with scheduling specifications.* Now we consider an STS $\Phi = \langle V, \Theta, \rho \rangle$ as before, but with the following modification for the transition relation $\rho = \rho(V^-, V)$:

$$\rho \subset \mathbf{dom}_V \times \mathbf{dom}_{\Lambda_V} \ , \tag{1}$$

i.e., transition relations relate the value for the t-uple of previous variables to the current state. As before, runs are sequences $s_0, s_1, s_2, \ldots$ that are consistent with transition relation (1).

STS involving such type of preorder relation will be called STS *with scheduling specifications.* STS with scheduling specifications are just like any other STS, hence they inherit their properties, in particular *they can be composed.*

Notations for scheduling specifications: for b a variable of type $bool \cup \{\perp\}$, and u, v variables of any type, the following generic conjunct will be used:

$$\textbf{if } b \textbf{ then } u \longrightarrow v \quad \text{, resp. } \textbf{if } b \textbf{ else } u \longrightarrow v$$

also written:

$$u \xrightarrow{\ b\ } v \text{ resp. } u \xrightarrow{\ \overline{b}\ } v$$

In [9], it is shown that scheduling specifications have the following properties:

$$x \xrightarrow{\ b\ } y \bigwedge y \xrightarrow{\ c\ } z \Rightarrow x \xrightarrow{\ b \wedge c\ } z \tag{2}$$

$$x \xrightarrow{\ b\ } y \bigwedge x \xrightarrow{\ c\ } y \Rightarrow x \xrightarrow{\ b \vee c\ } y \tag{3}$$

Properties (2,3) can be used to compute input/output abstractions of scheduling specifications. This is illustrated in figure 2. In this figure, the diagram on the left depicts a scheduling specification involving local variables. These are hidden in the diagram on the right, using rules (2,3).

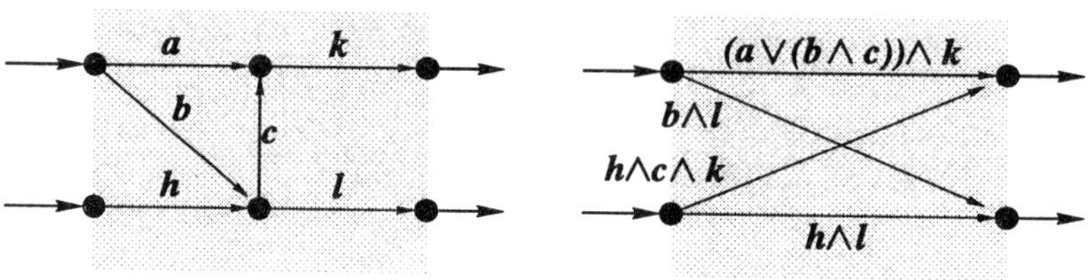

Fig. 2. *Input/output abstractions of scheduling specifications*

Syntax The following restricted set of generic basic conjuncts is sufficient to encode all known synchronous languages:

$$\begin{array}{c} \textbf{if } b \textbf{ then } w = u \textbf{ else } w = v \\ \\ u \xrightarrow{b} w \\ \\ \left.\begin{array}{rcl} w &=& f(u_1, \ldots, u_k) \\ h_w \equiv h_{u_1} &\equiv& \ldots \equiv h_{u_k} \end{array}\right\} \end{array} \qquad (4)$$

In addition to the set (4) of primitives, state-variable ξ_v associated with variable v can be used on the right hand side of each of the above primitive statements. The third primitive involves a conjunction of statements that are considered jointly.

Examples

$$\begin{array}{rl} \text{a selector}: & \textbf{if } b \textbf{ then } z = u \textbf{ else } z = v \\ \text{a register}: & \textbf{if } h_z \textbf{ then } v = \xi_z^- \textbf{ else } v = \bot \end{array}$$

$$\text{a synchronization constraint}: (b = \text{T}) \equiv (h_u = \text{T})$$

For the selector, the "**else**" part corresponds to the property "$[b = \text{F}] \vee [b = \bot]$". The more intuitive interpretation of the second statement is "$v_n = z_{n-1}$", where index "n" denotes the instants at which both v and z are present (their clocks are equal). Clearly, this models a register. This statement implies the equality of clocks: $h_z \equiv h_v$. The synchronization constraint means that the clock of u is the set of instants at which boolean variable b is *true*.

Inferring schedulings from causality analysis The schedulings that can be inferred from an STS specification result from 1/ explicit scheduling specifications, and 2/ dataflow dependences that result from causality analysis. The idea supporting causality analysis of an STS specification is quite simple. On the one hand, a transition relation involving only the types "pure" and "boolean" can be solved by unification and thus made executable. On the other hand, a transition relation involving arbitrary types is abstracted as term rewriting, encoded via directed graphs. For instance, relation $y = 2uv^2$ (involving, say, real types) is abstracted as $(u, v) \longrightarrow y$, since y can be substituted by expression $2uv^2$.

Scheduling specifications associated with the primitive statements 4 are given in [9]. For example:

$$\left.\begin{array}{rcl} w & = & f(u_1, \ldots, u_k) \\ h_w \equiv h_{u_1} & \equiv & \ldots \equiv h_{u_k} \end{array}\right\} \Rightarrow u_i \xrightarrow{\quad h_w \quad} w \tag{5}$$

Given an STS specified as the conjunction of a set of basic statements, for each conjunct we add the corresponding scheduling specification to the considered STS. This yields a new STS **sched**(P), for which it is possible to give sufficient conditions so that P is *executable:* roughly, **sched**(P) is provably circuitfree at each instant, and **sched**(P) has provably no multiple definition of variables at any instant. Then **sched**(P) provides (dynamic) scheduling specifications for the run of P.

2.2 Endochrony

The STS model is the semantical model of DC_+ programs. We need to give some sketch of the compilation of these programs in order to complete the internal representation structure of programs: one face is a directed graph, the other one is a *clock hierarchy* allowing to represent the control of the program. This structure is the basis for studying *endochrony.*

A DC_+ program describes a reactive system whose behavior along time is an infinite sequence of *instants* which represent reactions, triggered by external or internal events. The main objects manipulated by a DC_+ program are *flows,* which are sequence of values synchronized with a *clock.* A flow is a typed object which holds a value at each instant of its clock. The fact that a flow is currently absent is represented by the bottom symbol $\perp$ (cf. section 2.1). Clocks are pure or boolean flows. A clock has the value T if and only if the flow associated with the clock holds a value at the present instant of time. Actually, any expression *exp* in the language has its corresponding clock h_{exp} which indicates whether the value of the expression at the current instant is different from $\perp$.

Clock hierarchy. Directed graphs obtained by causality analysis such as presented above are one of the objects used to represent programs and to calculate on them. The other very important object that has to be obtained is a representation of the *clock hierarchy.*

A first step of the DC+ compilation is the construction of a system of equations on clocks. The system is solved, the result gives inclusion between clocks. The clocks are then organized in a *hierarchy* (i.e. a collection of trees) where if a clock h_1 is under a clock h_2 (called its father), then h_1 cannot be present if h_2 is not. In the clock hierarchy represented on figure 3, b, c denote boolean variables,

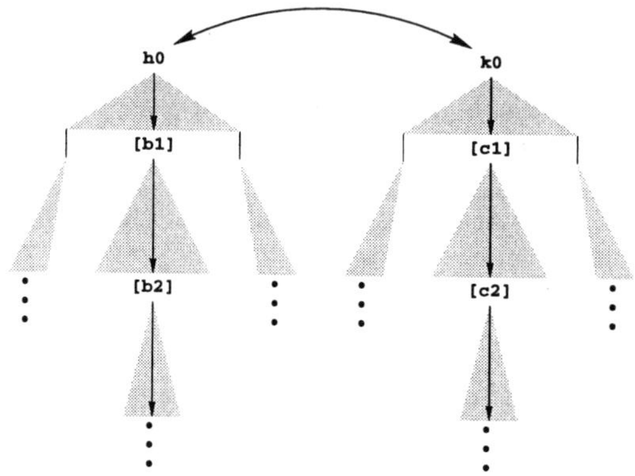

Fig. 3. *Clock hierarchy*

$[b], [c]$ denote corresponding clocks composed of the instants at which $b, c = \mathrm{T}$ holds, respectively. Finally, h, k are also clocks (i.e., variables of type pure). The down-arrows $h_0 \to [b_1]$, $[b_1] \to [b_2]$, etc., indicate that boolean variable b_1 has a clock equal to h_0 and only needs variables with clock h_0 for its evaluation, and so on. In doing so, a tree is built under each of the clocks $h_0, k_0, \ldots$, this yields the so-called *clock hierarchy* in the form of a "forest", i.e., a collection of trees. Roots of the trees are related by some clock equation, this is depicted as the bidirectional arrow relating $h_0, k_0, \ldots$ Then each flow of the program (and its definition) is attached to its clock in the hierarchy. This structure is detailed in [1] [2], where it is shown to be a canonical form for representing clock equations. The combination of clock equations and scheduling specifications of a program is represented by the combination of the clock hierarchy and of the directed graph.

Endochrony. An important property that will be determined on the clock hierarchy is *endochrony.* An STS is called *endochronous* if its control, i.e., the primary decision in executing a reaction, depends only on 1/ the previous state, and 2/ the values possibly carried by environment flows, but not on the presence/absence status of these flows. If an STS is not endochronous (it is *exochronous*), then the primary decision in executing a reaction consists in deciding upon relative presence/absence of clocks which are source nodes of the associated directed graph. In contrast, for an endochronous STS, only one *activation clock* is a source node of the graph. Hence no test for relative presence/absence of environment flows is needed.

It is shown in [9] that if a program P has a clock hierarchy consisting in a single tree, then it is endochronous.

Boolean DC_+. In practice, the calculation of the clock hierarchy of a program is a key tool of the DC_+ compiler; it is called the *clock calculus.* After the clock calculus, clocks can be defined as boolean flows in a sub-format of DC_+ called BDC_+: a boolean b represents the clock $[b]$ composed of the instants at which $b, c = \text{T}$ holds, and is defined itself at the clock which is the father of $[b]$ in the clock hierarchy.

2.3 Issues for modular and distributed code generation

Two major issues need to be considered for modular and distributed code generation:

1. *Relaxing synchrony* is needed if distribution over possibly asynchronous media is desired without paying the price for maintaining the strong synchrony hypothesis via costly protocols.
2. *Designing modules equipped with proper interfaces for subsequent reuse,* and generating a correct scheduling and communication protocol for these modules, is the key to modularity.

It is shown in [9] that a solution to the first issue is to restrict ourselves to endochronous programs. Another aspect that has to be considered is that of maintaining synchronous semantics of composition while using asynchronous communication media. Requirement for such a medium is that: 1/ it should not loose messages, and, 2/ it should not change the order of messages. The condition for this is the *isochrony* of the considered couple of programs. This is not detailed here, see [9].

The scheduling specifications we derive from causality analysis still exhibit maximum concurrency. Actual implementations will have to conform to these scheduling specifications. In general, they will exhibit less (and even sometimes no) concurrency, meaning that further sequentialization has been performed to generate code. Of course, this additional sequentialization can be the source of potential, otherwise unjustified, deadlock when the considered module is reused in the form of object code in some environment. We shall see in 4 that a careful use of the scheduling specifications of an STS will allow us to decompose it into modules that can be stored as object code for further reuse, whatever the actual environment and implementation architecture will be.

Enforcing endochrony. Since endochrony is a key feature for programs we have to implement, we must consider the question of moving from exochronous to endochronous programs. As shown above, an answer is to make the roots of the clock hierarchy belonging to some single clock tree. The idea for a simple example such as that of the figure 4 is to add to the considered STS a monitor which delivers the information of presence/absence via the b, b' boolean variables with identical clock h, i.e., $\{k = \text{T}\} \equiv \{b = \text{T}\}$, and similarly for k', b'. The resulting STS is endochronous, since boolean variables b, b' are scrutinized at the pace of activation clock h. Intuitively, moving from exochrony to endochrony corresponds

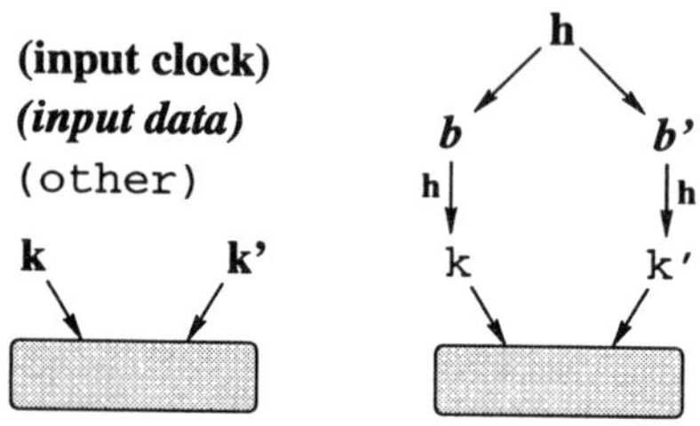

Fig. 4. *Enforcing endochrony*

to equipping the original P program with a suitable *communication protocol* Q in such a way that the compound program $P \bigwedge Q$ becomes endochronous. However, in the general case, this transformation is not unique [9].

3 Distributed code generation: overview of the SACRES approach

3.1 A first glance at the method

The overall method [6, 7, 16] is illustrated in the figures 5 and 6.

Figure 5 shows what the designer has to do. The designer has on her/his screen (at least) three windows (those of the top part of the figure). The first window—top left—is the (SIGNAL, or DC+, or Activity Charts in STATEMATE) program editor. In this window, a dataflow diagram is depicted. The arrows would typically depict flows of data, but also could correspond to scheduling requests. In the bottom right window, icons are shown which allow the designer to specify her/his target architecture. This architecture has two types of constitutive elements. The first one (on the left) is a *processor*, i.c. a machine that complies with the synchronous model of execution, in which a run is a sequence of atomic reactions. Thus processors can be pieces of sequential code (C/C++-procedures, threads, etc.), or alternatively parallel machines running according to the model of perfect synchrony (e.g., synchronous hardware). Other icons refer to (generally asynchronous) communication media. Using these two windows, the designer builds, in the third window (top right), her/his execution architecture: the source dataflow diagram is partitioned as shown in the figure, and corresponding subdiagrams are mapped onto "processors" by click-and-point. Also, models of communication links are specified by the designer, by clicking-and-pointing to the appropriate icon. The result of this mapping is built automatically as illustrated in the bottom part of the figure.

Figure 6 shows what the tool generates for each processor. From the specifica-

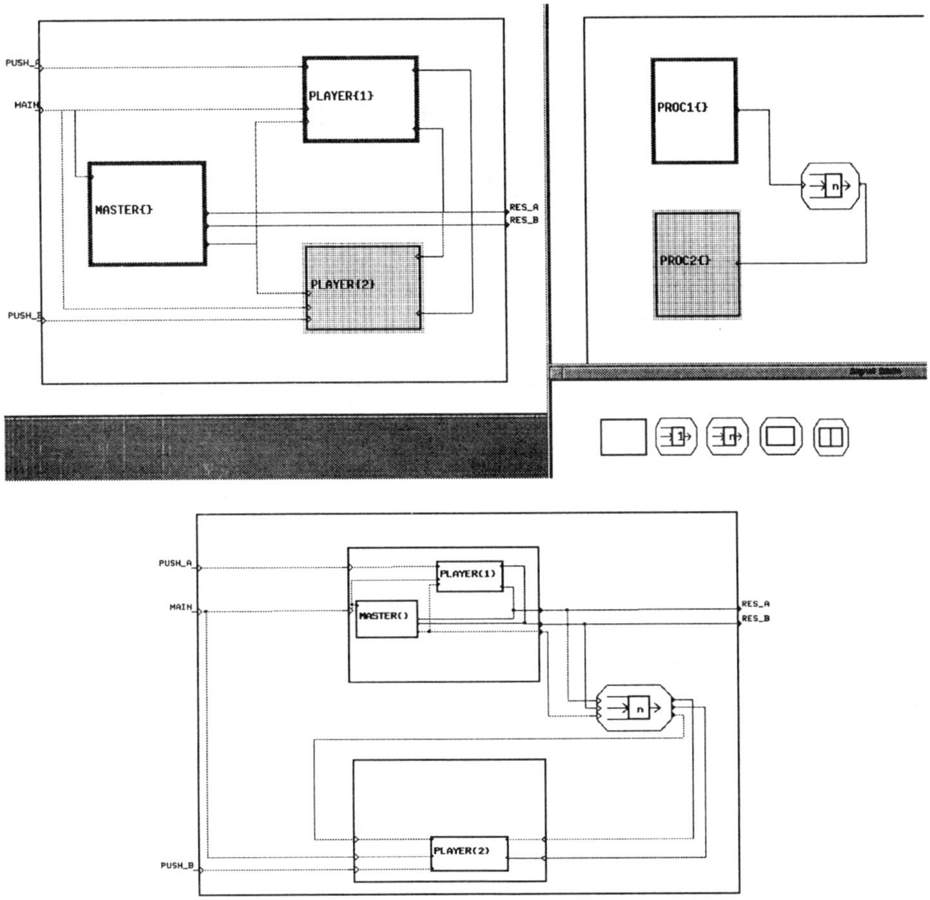

Fig. 5. *Code generation: what the designer does*

tions provided by the designer as in figure 5, the tool will generate, for embedding into each processor, the following: 1/ a suitable communication protocol which guarantees that the semantics of synchronous communication will be preserved even though an asynchronous communication medium is used; 2/ a structuring of the code into pieces of sequential code and a scheduler, aiming at guaranteing separate compilation and reuse.

Finally, the whole model (processors and channel models) can be used for architecture simulation and profiling.

3.2 Summary of the data needed to model the architecture

In the SACRES method, the path from the specification to the implementation goes through the DC_+ format. In this context, distributing an application consists in distributing a DC_+ program representing a functional graph of flows,

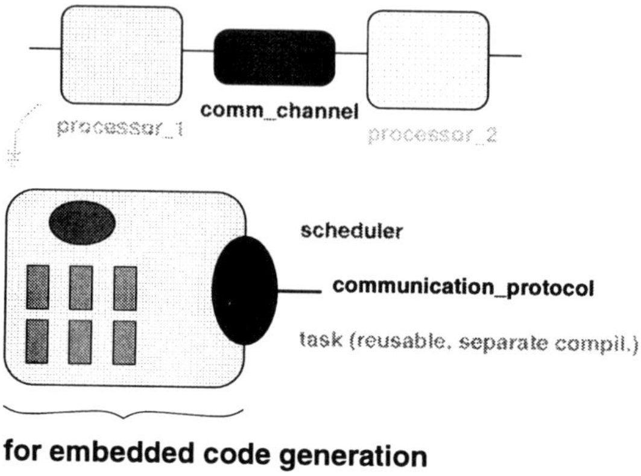

Fig. 6. *Code generation: what the tool generates*

operators and dependences. The target architecture is composed of a set of possibly heterogeneous set of execution components (processors, micro-controllers, ASIC...). At the level considered here, *processor* will be used as a generic name in the following.

We do not address the partitioning/scheduling problem against quantitative optimizations (as for traditional understanding in hardware/software partitioning). Conversely, we assume that an allocation function is given for nodes and flows to processors and links. Moreover, the user interface provided to describe this allocation function differs for specific environments (SILDEX, STATEMATE)—in the figure 5, the Inria environment for SIGNAL is used. These functions are ultimately given in pragma features of a DC_+ description.

We summarize here which data have to be provided by the users of architecture-dependent implementation of applications in the SACRES environment. A general comment is that the level of detail at which the architecture needs to be known depends quite a lot on the refinement of the mapping to the architecture chosen. This means that in the simplest cases, the amount of data required is fairly small, and simple to assess:

- the set of processors or tasks, and the mapping from operations or sub-processes in the application specification to those processors or tasks. This information enables the partitioning of the DC_+ graph into sub-graphs grouped according to the mapping.
- the topology of the network of processors, the set of connections between processors, and a mapping from inter-process communications to these communication links. This is useful in the case of flows exchanged between processes located on different processors or tasks, if several of them have to be routed through the same communication medium.
- a definition of the set of system-level primitives used e.g. for communications (readings and writings to the media). Roughly, this amounts to the profiles of the library of functions to which the code will have to be linked.

The concrete form to be given to this information is a question at two levels: the level of DC+ (which is where we actually perform the compilation work) and the user-level:

- the description of the location mapping can be made by having pragmas associated with nodes, defining their assigned location.
- the description of the architecture can be made using DC+, describing the graph of the network as a DC+ graph, with nodes for processors, nodes for communication links, edges for connections, and pragmas for attaching information to the nodes.
- at the user-level, tools and interfaces for entering this information are provided within the existing interfaces to SILDEX and STATEMATE.

The kind of information mentioned above supports the logical distribution of an application. In the perspective of having a more refined code generation, with further degrees of refinement of architecture-adaptation, more information has to be gathered on the architecture:

- concerning processors, taking into account the types of the different processors can lead to a code generation taking advantage of specific characteristics.
- concerning communications, the type and nature of the links (that could be implemented using shared variables, synchronous—blocking—or asynchronous communications...).

If the architecture targeted to features an OS, then in order to be able to generate code using its functionalities, the model needed consists basically in the profile of the corresponding functions, e.g., according to the degree of use of the OS, synchronization gates, communications (possibly including routing between processors) or tasking functions (in the case of un-interruptible tasks: starting and stopping; in the case of interruptible tasks: suspension and resuming, assignment and management of priority levels), etc.

4 Modular approach to the distributed code generation: main steps

When an application is executed on more than one processor, or in more than one task on a single processor, it is necessary to insure that the generated code is a correct implementation of the source specification. This correction has to be proved as far as possible. As long as the implementation process is some refinement in a single formally defined formalism, this proof results from theorems in this theory. Such an approach is adopted in the SACRES method.

We focus this presentation on synchronization and causality in the context of a structural decomposition of a synchronous program targeted on an optimized distributed code.

The code distribution is seen as a user activity allowed by providing a set of formally defined transformations using properties of the semantic model: commutativity, associativity, idempotence of the composition operator. The concept of static abstraction of a behavior (types of the flows plus their clocks and dependences) and parameterisation techniques make possible a modular approach to the distributed code generation.

Assumptions

- As mentioned above, we assume that a *locate* function gives the mapping from functional to physical architectures.
- We suppose the BDC$_+$ code of the program provided with a *well formed tree of clocks:* every flow x, including all the clocks but a single one (named *tick*) is associated with a boolean clock b_x. Each clock c is a boolean input or has an explicit circuitfree definition.
- Every flow which is not an input is functionally defined (no circular definition, the program is deterministic).

4.1 Virtual mapping

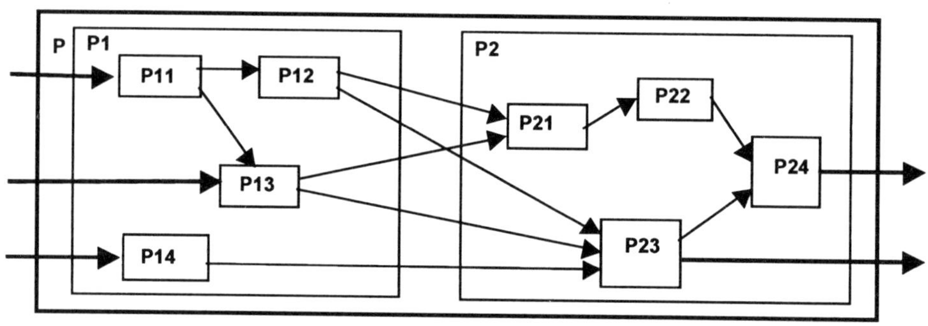

Fig. 7. $P = (P1 \bigwedge P2)$

Given:

- a source program $P = (P_1 \bigwedge P_2 \bigwedge \cdots \bigwedge P_n)$, in which each of the P_i may recursively be composed of sub processes $(P_{i1} \bigwedge P_{i2} \bigwedge \cdots \bigwedge P_{in})$, (in the example of figure 7, solid lines represent data-flow)
- a set of processors $q = \{q_1, q_2, ..., q_m\}$, and
- a function *locate:* $\{P_i\} \rightarrow \mathcal{P}(q)$ associating with each atomic process P_j a non empty subset of processors,

a process $Q = (Q_1 \bigwedge \cdots \bigwedge Q_m)$ is built (cf. figure 8); each process Q_i (source code to be implemented on the processor q_i) is the parallel composition of the set of atomic processes P_k such that q_i belongs to *locate*(P_k). It is easy to prove that Q is equivalent to P. As shown by the profile of the *locate* function, Q may include redundant processes (this is allowed by idempotence). We call *s-task* such a Q_i process.

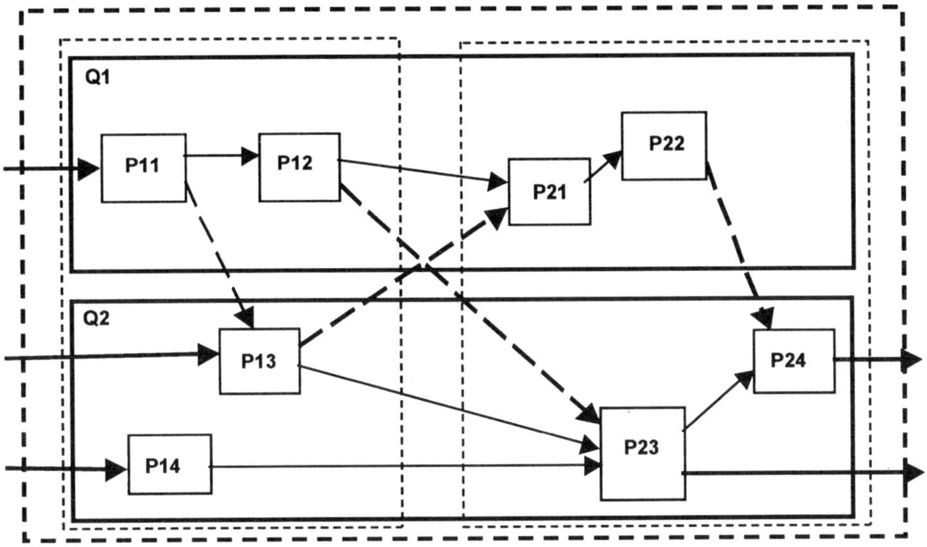

Fig. 8. *Virtual task allocation:* $P = (Q1 \bigwedge PQ2)$, *s-task* $Q1 = (P11 \bigwedge P12 \bigwedge P21 \bigwedge P22)$, *s-task* $Q2 = (P13 \bigwedge P14 \bigwedge P23 \bigwedge P24)$

4.2 Traceable compilation

In the process of distributed code generation, we have to care about the structure of the graph after restructuring. This graph is built in such a way that each sub-graph will be executed on a location. The compilation process must preserve this structure. As usual (cf. section 2.2) the first step of the compilation is the construction of a system of clock equations. Clocks are from the overall program. We next solve this system and build a clock hierarchy. The compilation of the virtual mapping is done without splitting the Q_i s-tasks; this compilation builds a global clock system and a global multi-graph.

Instead of splitting the graph across this hierarchy, we will project the hierarchy of clocks onto each sub-graph. Boolean clock definitions generated by the compiler are composed with the Q_i (located on processor q_i) with respect to some heuristic such that each one of the resulting Q'_i has a local tree of boolean clocks (BDC$_+$ endochronous code); the root of this tree is the upperbound of the clocks in Q'_i. We prune the clock hierarchy on each sub-graph to obtain a minimal one, but preserving a BDC$_+$ structure.

We finally prune the clock hierarchy of the graph of sub-graphs conserving only the fastest clock of each sub-graph.

4.3 Local interface abstraction

Each Q_i is associated with an interface containing:

- as inputs, the input flows of P (thick solid arrows in figure 8) and the flow computed in another Q_j, used in Q_i, (thick dashed arrows in figure 8)

- as outputs, the output flows of P and the flows computed in Q_i, and used in another Q_j,
- the clock tree of the input/output flows,
- the clocked dependences between external flows (inputs and outputs of Q_i). For that purpose, a transitive closure has to be calculated.

Transitive closure. Inside a sub-graph, a transitive closure allows to know precedence relations between input flows and output flows (we know if an output flow is preceded by an input flow); however we do not have any information on the other way around (i.e. between an output and an input, through the environment). The transitive closure on the whole graph is the only way to take a global view of the program.

In the case of a dataflow graph, the transitive closure is a well known algorithm. In the case of the DC_+ graph it is a little more complex one. We know that a dependence between flows is valid only at certain instants. All dependences of the transitive closure are also valid at certain instants and then associated with a clock. To obtain this clock, we have to apply rules (2,3) given in 2.1.

To avoid deadlocks at execution time, we have to add, in each sub-graph, precedence information resulting from the projection of the transitive closure on input and output flows. The code generation of a sub-graph must take into account these dependence relations to avoid making a dependence cycle when sub-graphs are executed together.

4.4 Local black box abstraction, sensitivity analysis

At the level of a s-task, we have to build a scheduling of the nodes of this s-task. The cost of dynamic execution leads us to reduce as much as possible dynamism. To do that, we will gather nodes is such a way that they can be considered in an atomic way. In this case, the scheduler only has to manage sets of nodes instead of nodes themselves.

Sensitivity equivalence

Definition *We say that two nodes N_1 and N_2 are sensitively equivalent if and only if for each input i:*
there is a causality path from i to N_1 $\Leftrightarrow$ there is a causality path from i to N_2.

Note that in this definition, we do not take clocks of dependences into account.

Two equivalent nodes wait for the same set of inputs: they can be executed in an atomic action depending upon this set of inputs. In this way, we build the classes of nodes transitively depending on subsets of input flows. For instance, the s-task Q_1 built above has two classes C_1 and C_2 as shown in figure 9.

Up to now, the applied transformations keep the program semantics unchanged.

Note that in the worst case, n input flows lead to 2^n sensitivity equivalence classes. In this case, the management of classes is almost as hard as the management of nodes themselves. Another way to do the partitioning into classes is to consider two or more input flows as atomic (considering two flows atomic is considering that they will be read or written at the same physical moment). Then the partitioning is done on subsets of atomic sets of input flows.

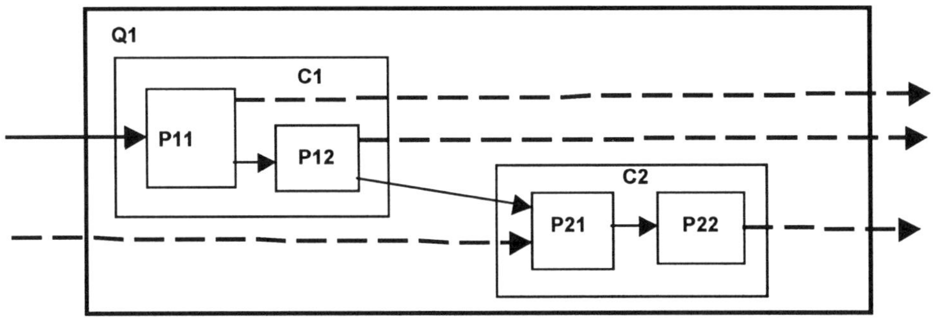

Fig. 9. *Sensitivity classes in* $Q1$

Local code generation All the nodes of a sensitivity class can be executed as soon as the subset of its input flows is available. We can generate any scheduling inside a class without modifying the observable behavior of the overall initial program. Thus, we will generate a sequential code, assuming that input values are available each time a new step starts for the class.

All the local flows, but state variables, can freely be implemented in registers or variables. State variables have to be implemented in remanent memory.

Black box abstraction We have said that inside a sensitivity class, we can generate any scheduling which respects dependences. This means that for any such class, the only dependences that we have to consider are that any input precedes any output at any time. For the sake of scheduling, a class is seen as a procedure call. The scheduling of classes in then done without taking into account what is inside, just the above property. For each class C_i, we just consider an interface in which all inputs precede all outputs. The abstraction of a class is a *black box* abstraction. For example, in the Q_1 sub-graph we obtain the result represented on figure 10, in which causality arcs have been substituted to the original dataflow sub-graphs C_1 and C_2.

4.5 Grey box abstraction for a s-task

If s-tasks are implemented as a single task on an execution entity, communication of internal data is done in the form of variable reads and writes in the same

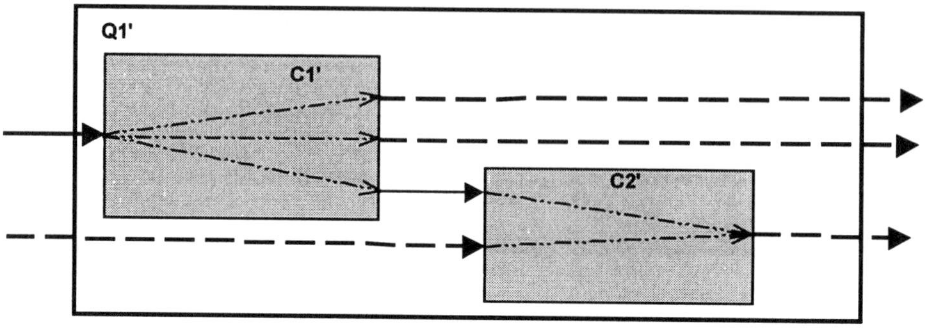

Fig. 10. *Sub-graph abstractions*

memory space. If s-tasks are implemented as a set of tasks, communication corresponds to inter-tasks communications as supported by the tasking system.

From the point of view of the environment, the kind of implementation chosen for a s-task does not matter. But to ensure the correct read-write sequencing, the internal communications have also to be abstracted as causality relations. Then we obtain what we call a *grey box abstraction,* in the sense that we know more than just its interface. For example, the grey box abstraction of the s-task Q_1 is depicted in figure 11.

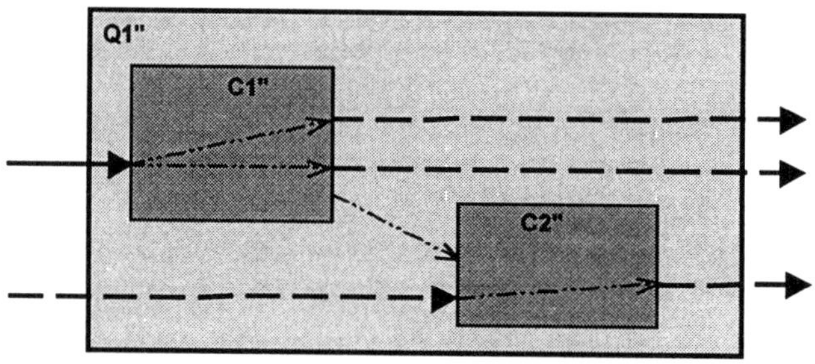

Fig. 11. *s-task abstraction*

4.6 Code generation for each s-task

Communications Inter s-task communication is communication between execution entities of an architecture. The communications are generated depending on the OS primitives. This generation is done by making calls to the right OS primitives given by the architecture description or by the user.

Communication features can be described as synchronous process abstractions (some *grey box*). We assume the mapping of links to these devices is given. Due to the large variety of communication features, multiplexing possibilities, synchronization relaxing policies, it is not possible to provide a general automatic mapping implementation; a specific library should be provided by the user (cf. figure 12).

To represent the new graph obtained by adding communications, we have to add causality links with the communication nodes.

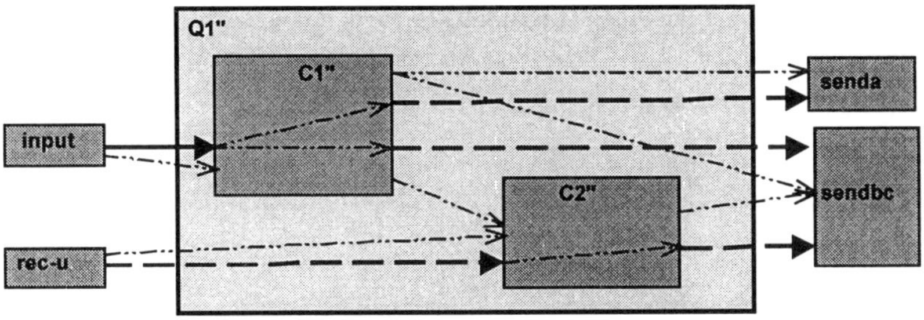

Fig. 12. *s-task abstraction with communications*

S-task scheduling A scheduling is computed for each s-task taking into account the global dependence graph and any other assertion on the environment.

In particular cases, for example obviously when the initial program is fully scheduled by construction, then we can build a static scheduling for its partition into s-tasks. In other cases, or when we want it, we have to build a dynamic scheduling, e.g. depending on the order of arrival of inputs of the different parts of the s-tasks. The order of the execution of nodes is given by scheduling clocks exchanged with the environment.

OS scheduling. For a task such as Q_1'' (cf. figure 12) it is possible to generate parallel processes scheduled and synchronized (synchronization gates) by OS primitives as shown for example in figure 13. In this case, nothing is known about the interleaving of elementary actions in T_1 and T_2.

Parameterized scheduling. Conversely, one can make more explicit the interactions between the s-task components and the OS; this is allowed by clocked causality arrows in DC+: considering two non ordered atomic actions A_1 and A_2 on the same processor, each time A_1 and A_2 are both executed, either A_1 is executed before A_2 or A_2 is executed before A_1; this variable scheduling can be represented by arrows from A_1 to A_2 labeled by S or conversely from A_2 to A_1 labeled by *not* S, where S is a boolean clock whose clock is the set of instants at which A_1 and A_2 are both executed. A single process is known by the OS. This is illustrated in figure 14.

The scheduling can then be described as a DC+ program depending on timing, values, etc., on which verification is possible; or also, S can be provided

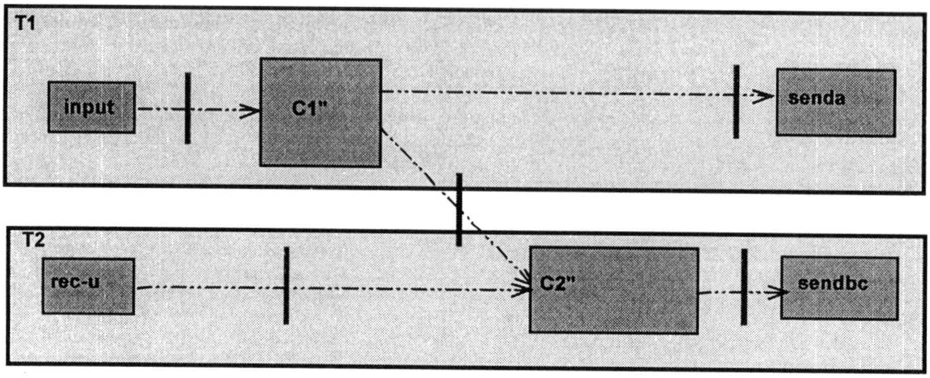

Fig. 13. *OS s-task scheduling*

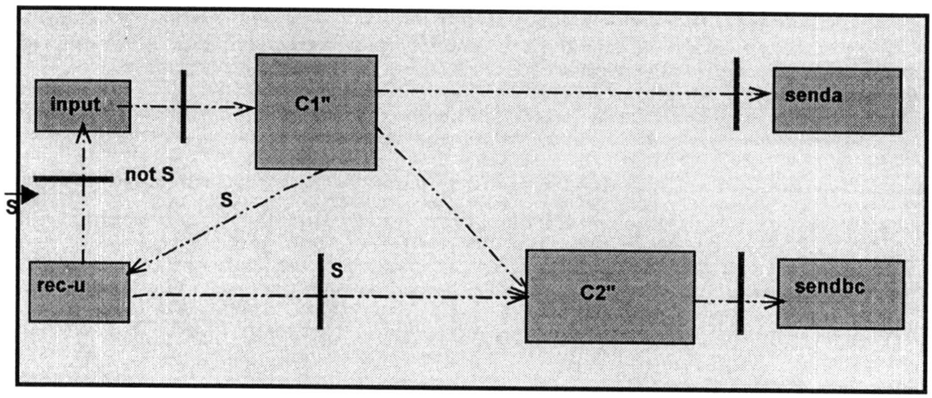

Fig. 14. *Parameterized s-task scheduling*

by the OS, or it can be a constant allowing then static scheduling as shown in figure 15.

Synchronization and communication The synchronization between tasks has to be defined by the user. Different schemes can be implemented.

Strong synchrony. The first implementation that can be done is to set a global gate to synchronize tasks. At each step, each activated task signals its completion to its predecessors in the graph when it has been completed and has received completion acknowledge from each of its successors. Thus at most one logical step is computed during the same instant.

Weak synchrony. A second scheme is to set local gates to synchronize tasks. At each step, each activated task signals to its predecessors in the graph that it is ready to proceed a new step when its previous step has been completed and a

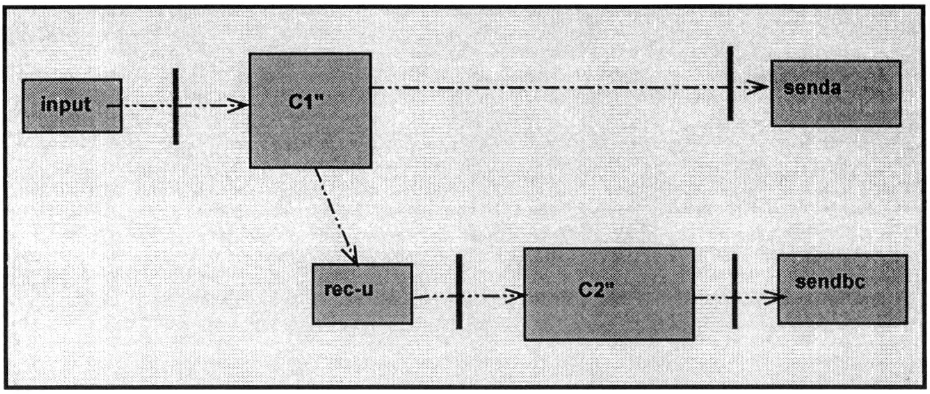

Fig. 15. *Static s-task scheduling with $S = true$*

"ready to proceed" has been received from each of its successors concerning this previous step. This implementation allows pipeline.

Bounded asynchrony. The most permissive scheme for embedded systems is to allow communication implementation as bounded FIFO; the *window* operator of DC_+ makes possible the semantic description of this partially asynchronous scheme.

4.7 Code instrumentation

Finally, we briefly mention our approach to code instrumentation, to evaluate e.g. performance [14].

Given an implementation Q of a program and a model of time consumption for each of the atomic actions in Q, we automatically generate a program $T(Q)$ homomorphic to Q; $T(Q)$ is the parallel composition of the images $T(Q_i)$ of the components Q_i (including communications) of Q. $T(Q_i)$ are given by the user as DC_+ components whose interfaces are composed of integer flows $T(x)$ instead of the original flows x. $T(x)$ represents the sequence of the availability dates for the occurrences of the original flow x.

$T(Q)$ is thus a model of real time consumption of the application (functional specification and architectural support). Some real time properties to be satisfied can be described as predicates in DC_+. Then these properties can be checked by using the verification tools of SACRES for instance.

5 Conclusion

This paper presents a framework for distributed code generation of synchronous programs that allows relaxing synchrony, thanks to the property of endochrony. On the other hand, the definition of precise abstractions of the programs permits reuse of separately compiled programs. So the method is also a method for separate compilation.

It is implemented in the SACRES project, through a number of software modules applicable to $\mathrm{D}\textsc{c}_+$ programs. These are for example:

- **clock calculus,** which is the core of the $\mathrm{D}\textsc{c}_+$ compiler and consists in computing the *clock hierarchy* (it is used in particular to check endochrony);
- **root adjunction, and event conversion,** implemented as a transformation of a $\mathrm{D}\textsc{c}_+$ program to a $\mathrm{BD}\textsc{c}_+$ one, which consist in inserting a *master clock* and converting clocks into boolean flows (this is used in particular to move from exochronous to endochronous programs);
- **building s-tasks,** based on user directives of location mapping, which performs the extraction of $\mathrm{D}\textsc{c}_+$ sub-programs;
- **computing abstractions of** $\mathrm{D}\textsc{c}_+$ **programs,** which consists in computing the transitive closure of dependences and projecting it onto the input/output interface;
- **building tasks,** which performs the extraction of tasks according to an input driven partitionning and calculates the scheduler of these tasks;
- **sequentializing** $\mathrm{D}\textsc{c}_+$ **programs,** which consists in preparing, for each executable program, the computing of a legal sequential scheduling;
- **distributed code generation,** performed on the result of the structuring of the code into tasks and s-tasks, which uses the sequential code generation for tasks and a specific code generation for the schedulers; the generated code makes calls to communication functions from a library to which it is linked.

The method can be applied to many possible targets, using different real-time kernels for instance (let us mention Posix, VxWorks, OSEK for automotive, ARINC for avionics...).

Also, thanks to the translation of STATEMATE to $\mathrm{D}\textsc{c}_+$ developed in the SACRES project [4,5], it is applicable not only to dataflow programs such as SIGNAL ones, but also to STATEMATE designs. Finally, it is partly available in the industrial SILDEX tool commercialised by TNI[1]. Application of the method has been carried out on case studies provided by industrial users [3].

References

1. T.P. Amagbegnon, L. Besnard, P. Le Guernic, *Arborescent canonical form of boolean expressions*, Inria Research Report RR-2290, June 1994, `http://www.inria.fr/RRRT/RR-2290.html`.
2. T.P. Amagbegnon, L. Besnard, P. Le Guernic, "Implementation of the dataflow language SIGNAL", in *Programming Languages Design and Implementation,* ACM, 163–173, 1995.
3. P. Baufreton, X. Méhaut, E. Rutten, "Embedded Systems in Avionics and the SACRES Approach", in *Proceedings of The 16th International Conference on Computer Safety, Reliability and Security, SAFECOMP'97,* York, United Kingdom, Springer, September 1997.

[1] http://www.tni.fr

4. J.-R. Beauvais, T. Gautier, P. Le Guernic, R. Houdebine, E. Rutten, "A translation of STATECHARTS into SIGNAL", in *Proceedings of the International Conference on Application of Concurrency to System Design (CSD'98)*, IEEE Publ., Aizu-Wakamatsu, Japan, March 23–26, 1998, 52–62.
5. J.-R. Beauvais, R. Houdebine, P. Le Guernic, E. Rutten, T. Gautier, *A Translation of Statecharts and Activitycharts into Signal Equations,* Inria Research Report RR-3397, April 1998, `http://www.inria.fr/RRRT/RR-3397.html`.
6. A. Benveniste (& contributors), "Safety Critical Embedded Systems Design: the SACRES approach", in *Proceedings of Formal Techniques in Real-Time and Fault Tolerant systems, FTRTFT'98,* Lecture Notes in Computer Science, Springer Verlag, September 1998.
7. A. Benveniste, T. Gautier, P. Le Guernic, E. Rutten, "Distributed code generation of dataflow synchronous programs: the SACRES approach", in *Proceedings of The Eleventh International Symposium on Languages for Intensional Programming, ISLIP'98,* Sun Microsystems, Menlo Park, California (USA), May 1998.
8. A. Benveniste, G. Berry, "Real-Time systems design and programming", *Another look at real-time programming,* special section of *Proc. of the IEEE,* vol. 9 n° 9, September 1991, 1270–1282.
9. A. Benveniste, P. Le Guernic, P. Aubry. *Compositionality in Dataflow Synchronous Languages: Specification & Code Generation,* Inria Research Report RR-3310, Nov. 1997, `http://www.inria.fr/RRRT/RR-3310.html`, see also a revised version co-authored with B. Caillaud, Mar. 1998.
10. A. Benveniste, P. Le Guernic, Y. Sorel, M. Sorine, "A denotational theory of reactive synchronous systems", *Information and Computation,* 99 n°2:192–230, 1992.
11. N. Halbwachs, *Synchronous programming of reactive systems,* Kluwer Academic Pub., 1993.
12. D. Harel, "Statecharts: A visual formalism for complex systems", *Science of Computer Programming,* 8:231–274, 1987.
13. D. Harel, A. Naamad, "The STATEMATE Semantics of Statecharts", *ACM Transactions on software engineering and methodology,* vol. 5, n° 4, 1996.
14. A.A. Kountouris, P. Le Guernic, "Profiling of SIGNAL Programs and its application in the timing evaluation of design implementations", in *Proceedings of the IEE Colloq. on HW-SW Cosynthesis for Reconfigurable Systems,* HP Labs, Bristol, UK, IEE, February 1996.
15. P. Le Guernic, T. Gautier, M. Le Borgne, C. Le Maire, "Programming real-time applications with SIGNAL", *Another look at real-time programming,* special section of *Proc. of the IEEE,* vol. 9, n° 9, September 1991, 1321–1336.
16. P. Le Guernic, S. Machard, E. Rutten, "Répartition de programmes SIGNAL", in *Actes des Rencontres Francophones du Parallélisme des Architectures et des Systèmes, RenPar'10,* Strasbourg, 9–12 juin 1998 *(in French).*
17. A. Pnueli, N. Shankar, E. Singerman, "Fair Synchronous Transition Systems and their Liveness Proofs", in *Proceedings of Formal Techniques in Real-Time and Fault Tolerant systems, FTRTFT'98,* Lecture Notes in Computer Science, Springer Verlag, September 1998.
18. SACRES consortium, *EP 20897 Deliverable report: The semantic foundations of* SACRES, March 1997.
19. SACRES consortium, *The Declarative Code DC+, Version 1.4,* Esprit project EP 20897: Sacres, Nov. 1997.

Formal Verification of an Avionics Application using Abstraction and Symbolic Model Checking*

Tom Bienmüller[1], Udo Brockmeyer[2], Werner Damm[2], Gert Döhmen[2], Claus Eßmann[2], Hans-Jürgen Holberg[2], Hardi Hungar[2], Bernhard Josko[2], Rainer Schlör[2], Gunnar Wittich[2], Hartmut Wittke[2], Geoffrey Clements[3], John Rowlands[3], and Eric Sefton[3]

[1] Carl von Ossietzky Universität Oldenburg, Germany
[2] OFFIS, Escherweg 2, 26121 Oldenburg, Germany
[3] British Aerospace, Warton Aerodrome, Preston, PR4 1AX Lancashire, UK

Abstract. This paper demonstrates the use of model-checking based verification technology to establish safety critical properties for an industrial avionics application. The verification technology is tightly integrated with the STATEMATE® system of i-Logix Inc., USA. Key features of this technology are its scalalability to complete system verification, the powerful debugging capabilities, graphical entry for safety critical properties, and the capability to re-use verification results for design components. The paper describes the application, the STATEMATE verification environment, and its use to establish safety critical properties of a British Aerospace application. The technical focus is on the use of abstraction techniques, allowing to focus verification on aspects of the design relevant to the property under investigation.

1 Introduction

This paper advocates the use of model-checking based verification technology to cope with the complexity challenge in designing electronic control units for safety critical applications. Model checking comprises an algorithm which takes a model of the system under design (SUD) and a (e.g. safety-critical) property, and checks—in a mathematically rigorous sense—that the SUD satisfies this property. From its basic algorithmic definition [CES83], model checking has matured to a state where it is on the edge of introduction into industrial design processes. Important milestones on this path were the introduction of symbolic representations of the SUD [McM93], the capability to focus on components of the design offered by compositional model checking [Jos93], the extensions to allow verification of complete systems by relying on properties of its components, and finally the use of abstraction techniques [CGL94]. Only the integration of these complementary approaches with an industry-standard design tool like STATEMATE

* This work was partially supported by the European Commission under Contract EP 20897 (SACRES)

makes the technology usable in an industrial setting. The industrial potential of the technology is demonstrated by the discovery of design errors which were not discovered by testing.

To assess the relevance of model-checking, it is instructive to view controller design of ECUs as a two-person game [BL69]: given the set of requirements, player one tries to set up a control strategy ensuring the requirements under all possible environmental conditions. The role of the enemy is played by the environment of the ECU: "attacks" are carried out by providing particular combinations of input stimuli over time, trying to drive the controller into a state violating the requirements. Examples of such attacks are "unexpected" sensor values, unanticipated combinations of sensor readings, or unanticipated sequences of combinations of sensor readings.

In this analogy, designing test sequences can be seen as inventing attacks. While in current design processes, this is dependent on the skill of the test staff, model checking can be seen as (implicitly) automating this process: when analyzing the SUD w.r.t. a given (safety-critical) requirement, the model checker takes into account all possible sensor readings as expressed by the type of the corresponding model input, all possible combinations of sensor readings, and all sequences of combinations of sensor readings. Hence, the model checker's analysis caters for all possible attacks. In contrast, no test set can be complete, and even a skilled test engineer may fail in finding the "critical" attacks simply due to the inherent combinational complexity of the SUD. In fact, errors discovered in the BAe application and other industrial trials share the property, that they require long sequences of stimuli with particular sensor settings across interfaces comprising often more than 50 inputs. How, then, can we possibly expect a test engineer to come up with such a particular attack?

While the above argument shows a qualitative superiority of model-checking based verification technology over classical test-based approaches, there is also a quantitative edge of model-checking technology over testing. To see why, view the SUD as an infinitely branching infinite tree, whose nodes correspond to the current state of the SUD, and whose branches depend on the combination of sensor readings in a particular state of the system. Applying a particular test sequence will follow one path of this infinite tree to finite depth. Model checking, to the contrary, explores in one sweep all attacks. In other words, one model-checker run conceptually replaces an infinite number of test runs. Based on this observation, model-checking technology has an industrial relevance not only in verifying safety-critical properties of a design—it also pays off to use the model checker early in the design process to debug components of the design. We will elaborate on use cases of the technology in the conclusion.

Given these advantages of model checking over testing, where are the drawbacks?

Model checking, in its pure essence as described in [CES83], is inherently a method for the analysis of finite state systems. Hence model variables with non-finite types must be abstracted into finite types for the method to be applicable in its current maturity level. Promising research [HGD95,BDG$^+$98] shows, that for many relevant application classes, even this restriction can be lifted, but these techniques are out of the scope of this paper. Assuming finite models, the immediate question popping up asks for a quantification. The break-through alluded to above [McM93] is the step from roughly a million states, i.e. 10^6 possible valuations of model inputs and variables, to, say, 10^{100} such valuations. However, this comes with a caveat: the encoding used is sensitive to model characteristics, and in fact one and the same model can have encodings greatly varying in size, depending on the ordering of the binary variables encoding the systems state [Bry92]. Even worse, there are clear cut theoretical arguments, why not all finite state machines can have succinct representations, and why finding the best encoding is in practice not feasible.

In spite of these discouraging news, model checking is alive and active. Reports on industrial applications abound—see e.g. the proceedings of the conferences on Computer Aided Verification—and products are on the market offering model-checking technology for ASIC verification. It turns out, that control-dominated, well-structured designs constitute a good match for the heuristics underlying the symbolic encodings and the employed analysis algorithms, and it is exactly this class of designs we find in safety-critical ECU designs. For such designs, it is safe to say on the basis of the large body of industrial applications, that models in the order of 10^{100} possible valuations can be handled. Moreover, the range of complementary techniques listed above—including compositional verification and abstraction—allows us to handle much larger systems, as long as the part of the design relevant to the property under consideration can be reduced to this size. As an example, the BAe case study breaks down to one thousand state bits, but still verification of a critical safety property is possible within minutes.

The work reported in this paper has been carried out within the SACRES project, which not only covers verification of specification models but as well automatic code generation of specification models and code validation techniques (c.f. also the companion paper reporting on automatic code generation [Ben98]). Combining the SACRES tool chain thus allows to automatically compile verified specification models to implementations, where the correctness of the compilation step is guaranteed by the code validation techniques [PSS98].

This paper is structured as follows. Section 2 gives an overview of the BAe application, demonstrating the complexity of the SUD, which will be used throughout the paper to illustrate the use of the various features of the STATEMATE verification environment. An overview of the functionality of the STATEMATE verification environment and its application to the British Aerospace SUD is given in Section 3. Section 4 focusses on abstraction and its relevance. The conclusion discusses possible use cases of verification technology in the design process.

Acknowledgements

The model checker underlying the STATEMATE verification environment has been developed at Siemens ZT, Munich, FRG. The realization of the STATEMATE verification environment was carried out in close cooperation with the team at Siemens ZT lead by Wolfram Buettner and Klaus Winkelmann, and with i-Logix Inc., Andover, USA. The semantic foundation was laid jointly with Amir Pnueli, Weizmann Institut of Sciences, Rehovoth, Israel [DJHP98].

2 The British Aerospace Application: A Stores Management System

This application developed by British Aerospace is a generic Stores Management System (SMS). The system is installed within the airframe and is responsible for the arming, pre-conditioning, release and jettison of all stores carried by an aircraft. The typical store types carried by a military aircraft are fuel tanks, bombs, rockets, missiles and practice bombs (and their carriers). These stores are connected to the airframe via a mechanism which is able, on command, to release the stores. The signal which commands the release is generated from the SMS which uses inputs from the pilot and other systems within the aircraft to control the release of these stores. A typical scenario is, if the aircraft is on ground, that the ground crew loads the stores, inputs the store configuration to the SMS and performs SMS ground checks. After that the aircraft takes off and armament power becomes available. When the aircraft approaches a target the pilot designates target using the Weapon Aiming Computer (WAC) and selects a store package (i. e. which stores are to be released) for release. As the aircraft nears the target, the pilot arms the stores and gives his consent to release the stores. When over the target, the WAC requests store release from the SMS and the SMS generates the store release signals. This is a general scenario which may change in detail if, for instance, the target is another aircraft. However, all normal store releases are generally of the same nature. The SMS has two main modes of release: "normal" and "jettison". The "normal" release was described above, the "jettison" release is used in emergency conditions and allows the pilot to release stores in an un-armed mode.

The application was designed with the aim of producing a working model of a Stores Management System that could be animated using the STATEMATE tool. In our description of the model, we assume some familiarity with STATEMATE. We will not explain such fundamental concepts as activity charts or state charts.

2.1 Hierarchical Tree of SMS

Each function of the SMS is expressed as an activity chart. Those are collected at system integration level in a single activity chart called SMS_ACTIVITIES.

The logical architecture of the system level refines into the subsystem architecture. In order to give a detailed overview of the subsystems of the decomposed system level of the STATEMATE design and to illustrate the data flow in SMS we present the functional decomposition in Figure 1. Informal descriptions of the functionality of each subsystem are listed below.

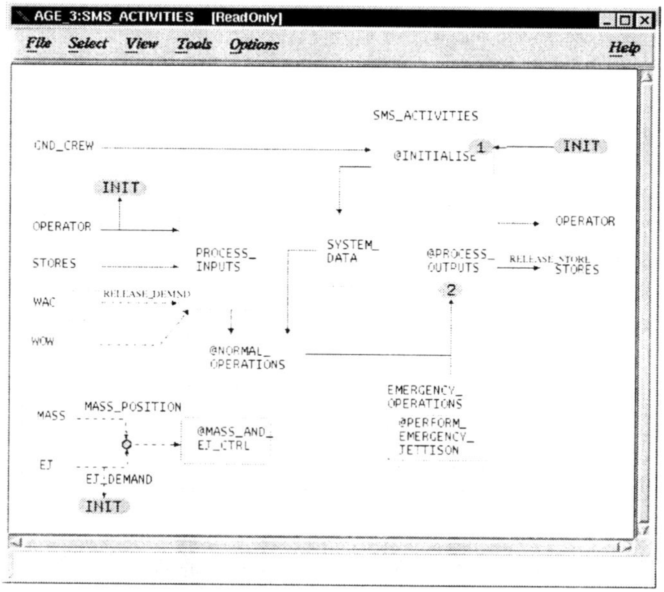

Fig. 1. Top-level activity (SMS)

MASS_AND_EJ_CTRL

This state chart is the embedded control unit of the Stores Management System. It implements the overall control over the other subsystems. The system power is controlled by the Master Arm Safety Switch (MASS) which has three positions: OFF, STANDBY and LIVE. When the MASS is OFF the SMS is unpowered and hence no operation is possible. When the MASS is in STANDBY the SMS is powered but no releases are allowed. Thus, in this mode the SMS completes any startup built-in test (not modelled in this case), resets any variables and sets all outputs to a non-active state. When the MASS is LIVE, the SMS remains powered and all functions including releases are allowed. When in this state, the SMS has two operating modes: LIVE and EMERGENCY JETTISON. LIVE is the normal case and the activity chart NORMAL_OPERATIONS is active. Operating the emergency jettison button will cause the SMS to enter the EMERGENCY JETTISON state which will deactivate the NORMAL_OPERATIONS activity and activate the EMERGENCY_OPERATIONS activity.

INITIALISE

This activity chart is simply concerned with initialising data at power-up and at the point of takeoff when the load-out panel data is copied to the inventory data. Because the SMS power can by cycled whilst the aircraft is in the air, a takeoff initialisation can be forced to happen. This will occur if the INITIALISE_ON_POWER_UP activity is activated whilst the aircraft is in the air.

EMERGENCY_OPERATIONS

This activity has been separated from the other activities in order to produce a functional split in the design. Because the emergency functions must be very robust it is hoped that this approach will aid with the verification of the emergency jettison function. In addition, the functionality of this activity has been kept as simple as possible. The purpose of this activity is to allow the operator to jettison all stores in an un-fuzed. Thus the activity provides the capability to lighten the aircraft under emergency conditions and to release any explosive material.

NORMAL_OPERATIONS

The majority of the SMS functionality resides within NORMAL_OPERATIONS. In here are held the activities that allow the operator to select release packages and to perform the release. The operator is allowed to define up to three release packages, each package consisting of a store type, quantity to release and the release sequence. The operator selects the store type and the quantity but it is the SMS which decides the release sequence. The release sequence is calculated such that the aircraft balance is maintained; in this case the aircraft will be balanced provided that the difference between the number of stores carried by the wings is at most one.

PROCESS_OUTPUT

This activity chart is concerned with generating all of the safety-critical outputs from the SMS. There are two sub-activities in this chart: DISPLAY_STORES which is concerned with all output to the operator and MAP_TO_PHYSICAL which is concerned with all store control signals. Because we want to prevent having multiple writer schemes across activity charts, the MAP_TO_PHYSICAL activity takes all station control signals from the different sources: NORMAL_OPERATIONS, EMERGENCY_OPERATIONS and MASS_AND_EJ_CTRL, and consolidates them into a single set of fuzing and release signals for each station.

PROCESS_INPUTS

This activity chart includes no STATEMATE functionality.

SYSTEM_DATA

This symbolises the global data memory for the inter-activity communication.

3 The Statemate Verification Environment

3.1 Overview

This subsection gives an overview of the key components attached to the STATEMATE system in order to support verification, as shown in Figure 2 on the following page. Figure 2 shows a central depository called verification data base, which

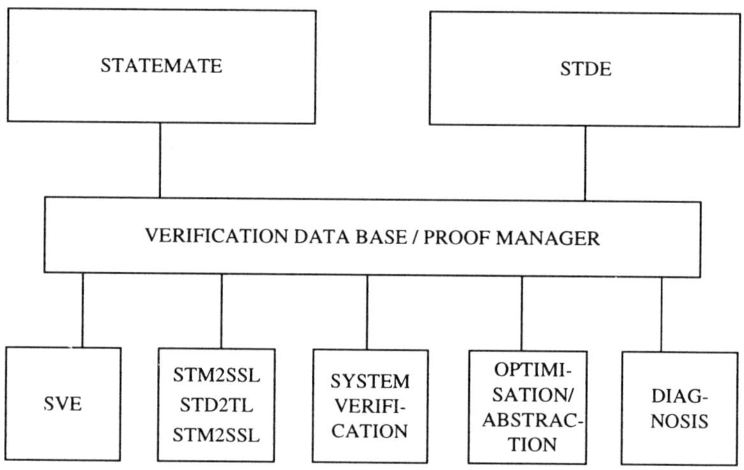

Fig. 2. The STATEMATE verification environment

maintains both models of components of the SUD as well as requirements on these. The kernel model-checking engine SVE developed at Siemens ZT accepts symbolic representations of the SUD in the FSM format, and temporal-logic requirements in the TL format. To capture requirements on STATEMATE models, a graphical entry tool STDE has been developed. STDE allows to express dependencies on changes of interface objects of the SUD. Three compilers provide the link between verification engines and the SUD as well as the requirement capture tool:

- STM2FSM translates STATEMATE models into symbolic transition systems; this complex compilation path is effectively split into two phases, passing through an intermediate format SMI discussed below; both the synchronous and asynchronous simulation semantics of STATEMATE are supported.
- STD2TL translates graphically expressed requirements into the temporal logic accepted by SVE.

- STM2SSL extracts from STATEMATE (a) for each activity an interface definition, and (b) a representation of the design hierarchy accepted by the system-verification and proof-maintenance tools; the acronym SSL stands for System Specification Language; the relevance of this interface is discussed below.

SMI stands for System Modeling Interface—an open standard allowing to attach modeling languages to verification technology developed at OFFIS [BBE98]. The current version of STM2SMI supports essentially all features of the STATEMATE system—see [Hol98] for a complete documentation. It must be noted, that unbounded types such as floats and queues can only be used in the verification process in connection with abstraction techniques—see Section 4. A number of optimisation tools are interfaced to SMI, allowing e.g. to reduce an SMI model w.r.t. the requirement to be verified.
SSL is used to represent design hierarchies and interfaces of multi-formalism designs independent of the employed modeling tool. SSL stands for System Specification Language [Döh97], and is in fact a sublanguage of VHDL [LSU95], inheriting in particular its capabilites of describing structures and configurations, as well as its package mechanisms.

When verification identifies an error situation, a powerful diagnostic component allows to analyse the cause of failure in two views. From a so called *error path* constructed from SVE, a trace of the SUD leading to the error situation can be visualized either as a timing diagram or compiled to a simulation script allowing to drive the STATEMATE simulator.
System verification allows to propagate properties established at the level of sub-activities upwards to the top-level chart, and is typically used, when a direct attempt to verify a top-level requirement with model checking failed due to complexity limits of the model-checking technology. Abstraction techniques allow to focus on relevant parts of the SUD, automatically reducing the complexity of its FSM representation, and thus effectively scale applicability of model-checking technology.

The following subsections highlight the usage of the components of the STATEMATE verification environment. To ease usage, a proof manager [Win98] has been developed, which allows to construct verification tasks of the form "does activity A match requirement R" for any of the (sub-)activities of the SUD and requirements captured through STDE. The appropriate calls for establishing the verification task are invoked automatically by the proof manager.

3.2 Requirement Capture with STDE

Like state charts, symbolic timing diagrams (STD) [FJ97] form a so-called visual formalism, which means that it has graphical objects as primary building blocks with a formal semantics ascribed to them.

While the graphical formalisms used in the STATEMATE system are tailored

towards a concise description of reactive systems based on an operational paradigm, STD takes a complementary role: The formalism is mainly used to describe certain critical properties of a system under design in terms of its input/output behaviour. The semantics is based on a declarative paradigm: STD specifications use sets of diagrams, where each diagram describes one aspect of the required input/output behaviour independently. The semantics of a STD-specification is a direct function of the semantics of the individual diagrams in the set.
This is deeply related to a particular methodology of system verification. The construction of a requirement specification in terms of several (conjunctive) specification clauses leads naturally to a high degree of modularization. The verification of a full specification consists of the verification of all diagrams (specification clauses) in the set. The idea to look at one requirement at a time both helps in terms of verification complexity and in terms of specification validation. If a particular property (diagram) is violated, it is much easier to compare the resulting error trace against a single diagram rather than a set of diagrams or an otherwise compound specification.

In this paragraph we give five example properties of the Stores Management System, specified by symbolic timing diagrams. At this integration level the specifications range over the observables of the subsystems, e.g. the "confirmation" stipulated by the system requirement can be specified by events of the observables LATE_ARM and RELEASE_CONSENT of the subsystem level. Examples of informal subsystem requirements derived from system-level requirements are:

1. Can we prove that except under emergency jettison condition releases only occur on a release-demand signal?
2. Can we prove that except under emergency jettison condition LATE_ARM and RELEASE_CONSENT are active during the release signal?
3. Can we prove that the interval between store releases under emergency jettison condition will never be less than the Minimum Release Interval (MRI)? The MRI is set to 10ms.
4. Can we prove that the interval between store releases under emergency jettison condition will never be longer than the Maximum Release Interval (MaxRI)? The MaxRI is set to 100ms.
5. There is a hard-coded Emergency Jettison Sequence (EJS). Can we prove that this is the only release sequence under emergency jettison condition? The EJS is set to (1,5,2,4,3) if we have 5 stores on the plane.

Considering these example requirements we distinguish the aspects of the different design-views concerned by the requirements. Requirement (1) does not constrain performance aspects of the system, but stipulates a safety-critical property. In contrast, properties (3) and (4) express requirements of system response time. Hence they are related to the performance-related design view. The last property rather falls into the category of a functional design view because it neither constraints performance nor does it express safety-critical requirements in that sense.

Consider first the example requirement (1):

Can we prove that except under emergency jettison condition releases only occur on a release-demand signal?

Figure 3 shows the formalization of this requirement. It states that a release of a store is performed only if it has been demanded by the pilot. Performing a re-

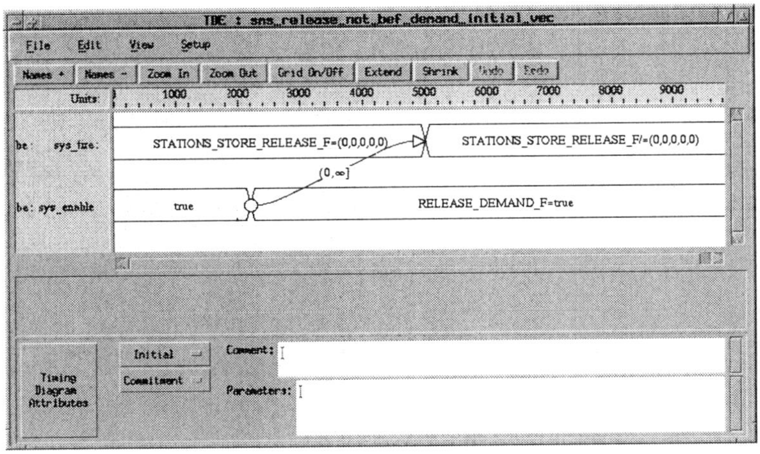

Fig. 3. Timing Diagram specifying that releases only occur on demand

lease is observable by a change on the signal STATIONS_STORE_RELEASE_F. Each component of this array contains a flag indicating a release with respect to a certain store. For example, STATIONS_STORE_RELEASE_F[2] indicates the release state of store 2. If it is set to '1', then a release is performed, if it is set to '0', then no release is done for that store. A change of the signal's valuation from a vector of '0's—indicating that no store is currently being released—to a state where some components of STATIONS_STORE_RELEASE_F are set to '1' indicates that a release command is issued by the SMS. The second signal required for specification is RELEASE_DEMAND_F. If this signal is set to 'true', then a release has been demanded. If this signal is set to 'false', then no release is allowed[1]. By linking both observations by an arc as shown in Figure 3 it has been stated that a release of any store is allowed to happen only if it was demanded.

The second property can be specified by a symbolic timing diagram (Figure 4) with one waveform. This diagram specifies an invariant for the subsystem which should hold over all system runs. In our case the diagram expresses that if a release occurs, LATE_ARM and RELEASE_CONSENT are active.

Requirements (3), (4) and (5) can be expressed in one initial real-time timing

[1] This description does not take into account releases due to emergency jettisons. This case has been excluded by an assumption.

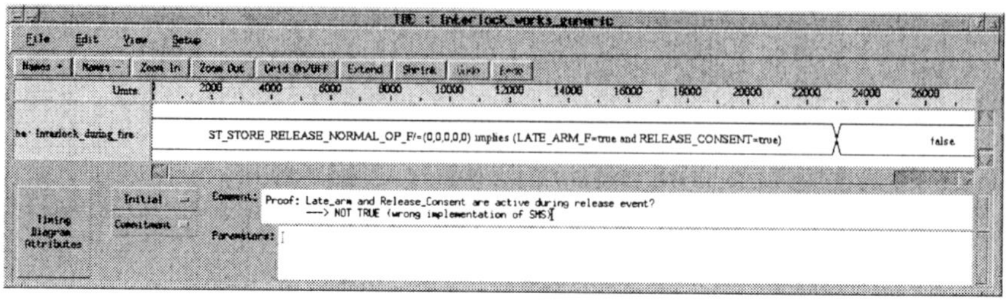

Fig. 4. Property (2) as an initial timing diagram

diagram presented in Figure 5. All timing constraints in Figure 5 are annotated with timing intervals. The numbers are representing the numbers of steps in the STATEMATE step semantics (synchronous simulation semantics) and each step is defined to be 5 ms. The constraints between the store releases are real-time constraints within the interval from 10 to 35 ms. Hence these real-time constraints match the real-time properties (3) and (4). Property (5) could be specified by using the same diagram as in Figure 5 without real-time intervals, thus only constrained by symbolic constraints with annotations (0,?], but is also covered by the diagram of Figure 5. It is important to know that our notion of real-time

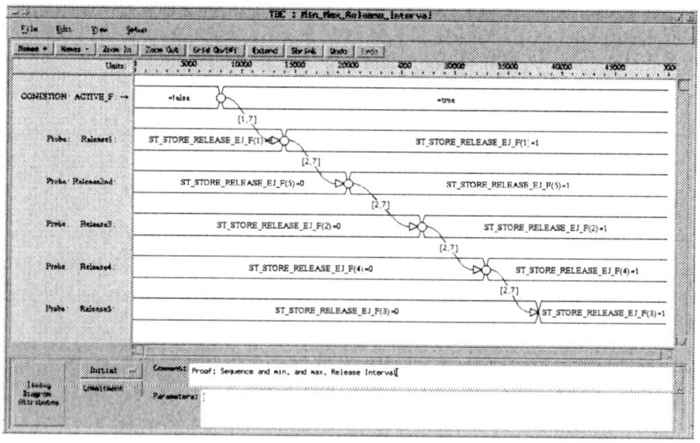

Fig. 5. Real-time timing diagram for properties (3), (4), and (5)

model checking is the notion of counting steps with a specified, fixed duration. The actual duration of a step in an implemented system depends on the executing target machine.

The above examples were intended to give a flavour of the graphical specification style of requirements supported in the STATEMATE verification environment. The interested reader is referred to the following publications for a deeper understanding of the semantics of STD: [DJS95] was the first complete reference to the semantics of STD with logical constraints only. The semantics for the extension of STD allowing timing constraints is described in [FJ97]. For a more user-oriented description of STD, the tutorial [Jos97] is most useful.

3.3 Applying Model Checking

To apply model checking, the designer uses the familiar concept of compilation profiles to generate an SSL/SMI representation of the selected subdesign of the SUD. Currently, the designer is required to then explicitly make the design visible in the verification data base, and hence to the verification tool set. He can then invoke the proof manager, which allows through incorporated browsing facilities to describe verification tasks by simply selecting the desired components of the SUD and the requirements to be checked. Flags controlling various optimisations and heuristics in particular pertinent to dynamically recomputing the "best" encoding can be controlled through panels.
For the complete SMS system configured with 5 stores the concrete model consists of 634 state and 68 input bits—the base version of SMS equipped with 12 stores needs 1152 state and 85 input bits to encode the behaviour of the model. The subactivities NORMAL_OPERATIONS and PERFORM_EMERGENCY_JETTISON require 376, resp. 52, state bits and 98, resp. 7, input bits in the five-stores configuration. Compiling these down to FSM level takes around 3000 seconds for the full SMS, and 500 seconds, resp. 10 seconds, for its subactivities.

Without the abstraction features available only recently, it was not possible to verify the top-level safety requirement on the full SMS system using model checking. Instead, the system verification capabilities have been invoked, as described in Section 3.5. All properties (1) to (5) could either be verified (properties (1),(3),(4),(5)) or falsified (property (2)) on NORMAL_OPERATIONS, resp. PERFORM_EMERGENCY_JETTISON, taking around 2000 seconds for each requirement on NORMAL_OPERATIONS, and less than 10 seconds on PERFORM_EMERGENCY_JETTISON.

3.4 Error Diagnostics

One of the most powerful features of the STATEMATE verification environment is the diagnostic support offered when model checking identifies an inconsistency between requirement and model. In such a situation, an "attack" is constructed, driving the contoller into the error situation. The tool set allows to either visualize this attack as a timing diagram, showing the sequence of input valuations leading to control failure, or to animate the STATEMATE model with a driver encoding the attack.

Consider the example requirement (2):

Can we prove that except under emergency jettison condition,
LATE_ARM and RELEASE_CONSENT are active during the release signal?

The verification of property (2) shows that NORMAL_OPERATION does not satisfy this specification. Figure 6 shows the resulting error-path visualisation in a "simple" timing diagram. Each dashed line over the waveforms symbolizes a step boundary resp. to the synchronous step semantics [HN95]. This error path consists of 8 steps until the violation of our property condition is detected. During the weapon release event of store 1 (in Figure 6 at x-position 80000 (step 8) [0,0,0,0,0] ([1,0,0,0,0]) the interlocks LATE_ARM and RELEASE_CONSENT are *not* active (false). So this execution sequence of the component is a counterexample to property (1) of NORMAL_OPERATION.

In the waveform "NORMAL_OPERATIONS_END_PERFORM_RELEASE"

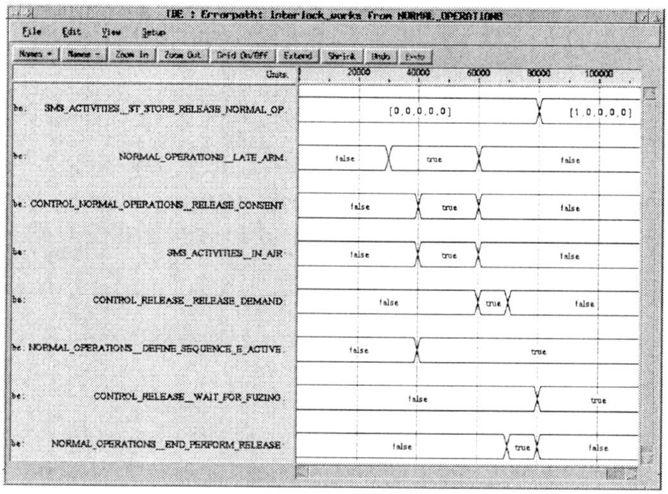

Fig. 6. Error-path visualisation of property (2) on entity NORMAL_OPERATION

we can see that a release stop signal (the meaning of this event variable) is generated in step 7 after the falling edge of the interlocks (NORMAL_OPERATIONS_LATE_ARM and CONTROL_NORMAL_OPERATIONS_RELEASE_CONSENT) in step 6 inside of the model of NORMAL_OPERATION, but there is no reaction of the model to prevent the release event (Waveform: SMS_ACTIVITIES_ST_STORE_RELEASE_NORMAL_OP) in step 8. This observation may be sufficient for the designer to spot the error in the STATEMATE implementation, knowing where the release stop signal is generated.

If, however, the designer wants to see the run in a more design-related manner, he/she can drive the STATEMATE simulator with a simulation program (SCD) corresponding to the error path. During the simulation of the error path, it is possible to observe every design element like data items, state charts, activity charts, etc.

Figure 7 shows a snapshot of the simulation of the attack at step 6.

In the right-hand window the simulator shows that every sub-component is

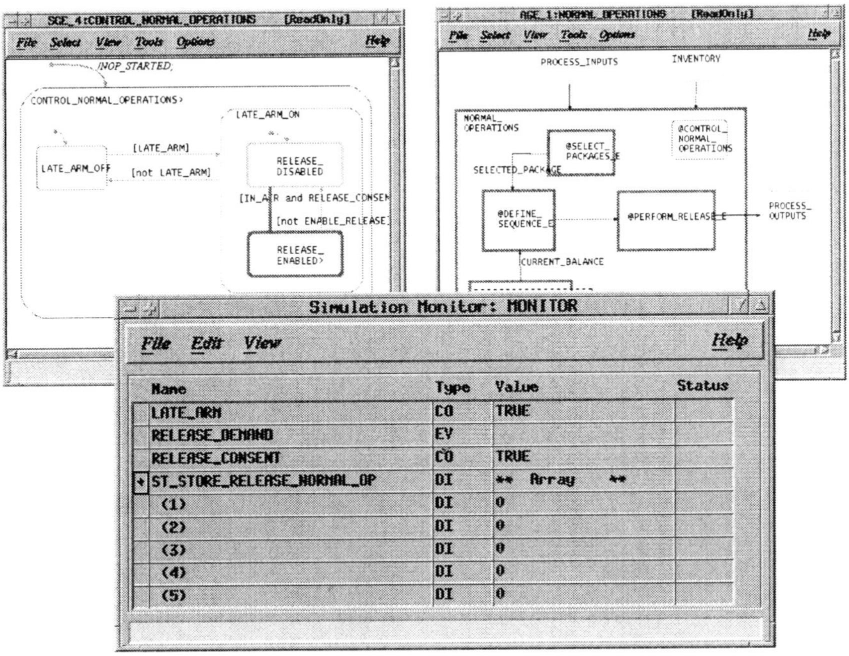

Fig. 7. Error-path simulation of property (2) in step 6

activated by the controller CONTROL_NORMAL_OPERATIONS in step 6. The controller (Figure 7, right picture) is in the state RELEASE_ENABLE, because the interlocks (LATE_ARM and RELEASE_CONSENT) are active (see window at the bottom of Figure 7).

Step 8 shows that the weapon release event is generated by the model (STNCTRL01: right picture of Figure 8) while the interlocks are not active (monitor window in picture of Figure 8). The reason is, that the chart CONTROL_RELASE (left picture of Figure 8) does not terminate directly after the END_PERFORM_RELEASE event from the controller. The chart CONTROL_RELASE can enter the state WAIT_FOR_FUZING and terminate later.

3.5 System Verification

When—even after applying abstraction techniques—the SUD as a whole cannot be analyzed via model checking, the tool set allows to derive top-level safety properties based on properties of its subactivities, hence avoiding the need to

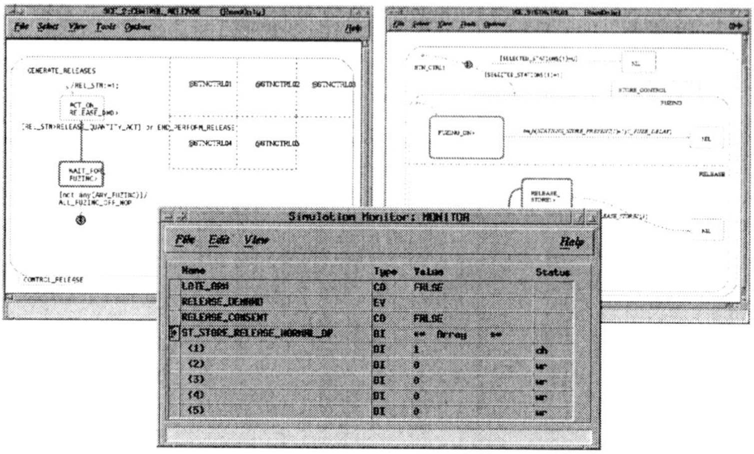

Fig. 8. Error-path Simulation of property (2) in step 8

construct one FSM for the complete model. This proof principle is often called compositional verification, since the verification of the complete system is based on the verification of properties of its components, using either model checking or again compositional verification.

Consider again the safety property (1) of the SMS design

(A1) Except under emergency jettison conditions,
(C) no stores are released, unless a release demand occurs (from the WAC)

From the formulation of the property, a strategy for logical decomposition can be derived. The clause (A1) “Except ... ” is an assumption of the requirement expressed in the clause (C). Assumptions restrict the set of all possible system behaviours. With respect to the distribution of the functionality in the system, global assumptions may effectively deactivate certain parts of the system. In our particular example, only one data path remains critical to the requirement (C) under the assumption (A1).

Thus, the strategy for logical decomposition of the global property to component properties is to trace back the requirement along the critical data path (see Figure 9) to find the “main cause” (technically, the relevant system component), which ensures that the requirement is in fact a valid property of the full system. In our case, the main cause why the property is valid for the full system is that there is a particular component, which “generates” the releases in normal operation mode. Although there are other potential sources for the generation of releases (in particular the “emergency jettison” mode), these are excluded by the global assumption (A1).

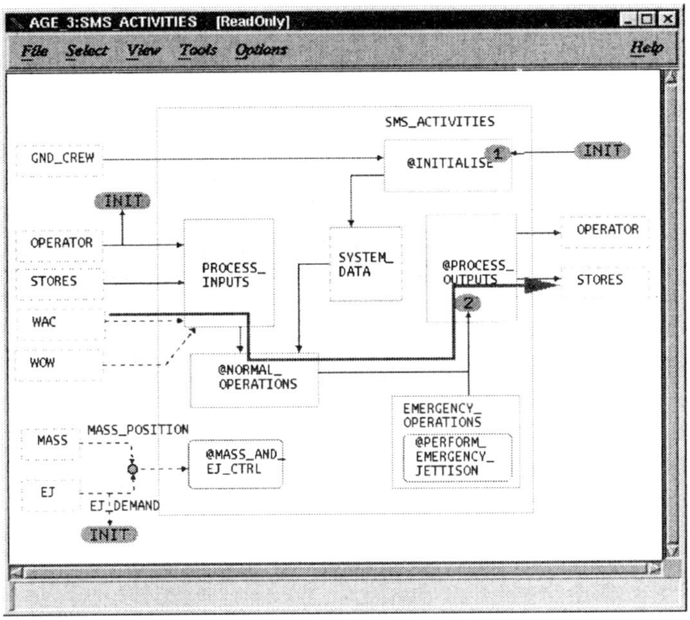

Fig. 9. Critical data path

Planning the proof: Bottom-up vs. top-down.

The discussion in the previous subsection suggests a top-down approach: Starting from an analysis of the system requirement, local requirements are formulated along the given system structure. Whether the local requirements are sufficiently strong to guarantee the global proof goal can be decided using an automatic method called *tautology checking*, essentially implemented by just calling the model checker in a different mode. If the logical decomposition is valid, then this is confirmed by the method. Otherwise, a counterexample is generated which helps to understand why the premises are too weak, and which can be visualized as a timing diagram.

Clearly a decomposition of a global property into local properties is useless, if the sub-activities do not satisfy the local requirements. This motivates to work also bottom-up: Given a thorough understanding not only of the full system as a whole, but also of its individual components, one should provide "grey-box" characterizations of the component interfaces. Given these characterizations, it will be possible to derive any valid global requirement, provided the component behaviours have been characterized in sufficient detail.

Such a grey-box characterization in general helps to support re-use of verified components, and also provides a means for IP-protection: rather than providing complete models, integration can be based on the cahracterization of the component wrt timing, functionality, safety, and protocols for inter-task communication and synchronisation made available in the grey-box view.

In practice, the verification work will employ a combination of a top-down and a bottom-up approach. Using system verification, the top-level safety requirement (1) shown in Figure 3 has been established for the SMS system.

4 Using Abstraction to Verify Safety-Critical Embedded Systems

While compositional reasoning requires the manual activity of decomposing global properties into local component properties, ***abstraction techniques*** can be applied to reduce computational complexity and hence allow verification of complete systems in a fully automatic way.
Several techniques like data abtractions [CGL94] and abstractions based upon data-independence assumptions [Wol86] have been introduced in the literature. Some of these techniques require a high degree of user interaction: i. e. the verification engineer has to supply mappings between concrete and abstract state spaces [CGL94]. This section of the paper focuses on an easy-to-use abstraction technique and its application methodology. This abstraction has been embedded into the STATEMATE verification environment.

A model checker verifies automatically whether a model fulfills a given requirement. Most properties to be checked this way refer to just a fairly small part of the model. By a dataflow analysis, one can compute the *cone of influence* with respect to a given specification $\mathcal{S}$, which contains all parts of the system which might possibly influence validity of $\mathcal{S}$. I.e., one can prove or refute $\mathcal{S}$ with an FSM representing only that cone. Figure 10 shows the cone of influence of the SMS model for requirement (1) referring only to one output (in Figure 10 depicted by the arrow pointing to "STORES")[1]. In this figure components outside the area marked by the dashed lines do not influence the outputs referred in the requirement.

In the STATEMATE verification environment, the cone of influence is calculated automatically as part of the optimisation tool [Ess98] on SMI level. Abstractions achieve a further reduction of FSM complexity, permitting the checking of still larger designs.

For each cone of influence $\mathcal{K}^{\mathcal{S}}_{\mathcal{M}}$ of a model $\mathcal{M}$ and requirement $\mathcal{S}$ there is a set $\mathcal{V}^{\mathcal{K}^{\mathcal{S}}_{\mathcal{M}}}$ of variables describing the model's state space. Each of these variables is associated with a function specifying the next-step relation between two states. For many properties to be verified on $\mathcal{M}$, there is no need to know concrete values of certain variables.

As an example, let us consider the safety property which interlocks releases of

[1] Note that the cone of influence contains the critical path from Figure 9, described in Section 3.5.

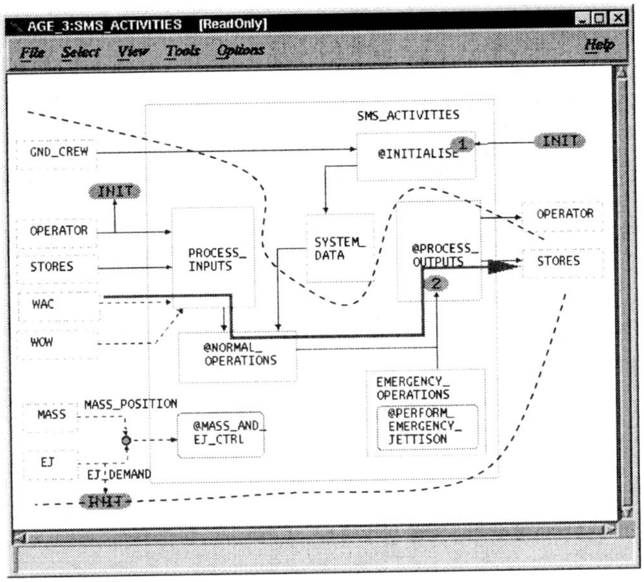

Fig. 10. Cone of Influence of the SMS model w.r.t. to a requirement referring only to outputs to "STORES"

stores with the explicit consent of the pilot, see Figure 3. Clearly, satisfaction of this property will not depend on the particular sequence of releases, though it is in the cone of influence. Hence, for this property, we can abstract from the computation of the concrete release sequence. We call this process "freeing" the variable maintaining the release order, entailing that in the abstracted model any of its possible values will be considered. In particular, in the FSM representation there is no need to "remember" exactly how this sequence was constructed—abstraction by freeing a variable effectively turns this variable into an input of the FSM analyzed in model checking, hence model checking will consider all possible valuations when constructing attacks.

The STATEMATE verification environment supports construction of abstract models for this kind of abstraction in three different kinds of granularity. In this paper, we omit discussion of their theoretical foundation and focus on their effects on verification.

Let `a` be a variable to be freed.

- **coarse grained:** at the beginning of each step, the value of `a` is set to an arbitrary value. Within the step, the computation refers to this "guessed" value.
- **middle grained:** each occurrence of `a` within terms of the model is substituted by a fresh input. For example, an assignment

```
x := a + b
```

will be converted to

$$\texttt{x := } a_{inp} \texttt{ + b,}$$ [1]

where a_{inp} is a fresh input with type of a.

- **fine grained:** whereas in middle-grained abstraction some information about the effects of variable a remains in the model, in fine-grained abstractions this information will be totally lost. For example the assignment shown above will be converted to

$$\texttt{x := } x_{inp},$$ [2]

where x_{inp} is a fresh input with type of x.

There are some more effects of the chosen abstraction type like their influence on conditions etc—these will be ommitted here.

Whereas the *coarse* type just turns variables into inputs, the *middle* type inserts more than one fresh input per variable to the model. In this kind of abstraction, each occurrence of a variable to be freed is substituted by a fresh input. It is easy to see that nondeterminism increases when applying the middle-grained abstraction type, and that several inputs for one variable may become visible in an error path. The *fine*-grained abstraction increases nondeterminism even more, since also variables which are influenced directly by an abstract variable get arbitrary values each step. While fine-grained abstractions lead to smaller models, coarse-grained abstractions allow better readability of error paths.

4.1 Application Methodology

As described in the previous section, abstractions of this type require identification of an appropriate set $\mathcal{A}$ of variables, which also occur in the cone of influence with respect to a model $\mathcal{M}$ and a requirement $\mathcal{S}$, but are independent of the property to be verified. Recall, that variables outside the cone are eliminated automatically beforehand. To select this set $\mathcal{A}$ is a creative task and has to be performed by someone familiar with the model and the specification.

Once $\mathcal{A}$ has been selected, the abstract model can be derived from the concrete one without further manual intervention.

Regarding applicability of abstractions it is important to note, that counterexamples generated by the model checker stay comprehensible, i.e. if $\mathcal{A}$ is selected carefully, then the user is able to distinguish between "bad" abstractions and errors in the model or the specification. Thus, also a negative verification results provide information about the concrete system.

The selection of the abstraction type depends on the user's "familiarity" with

[1] Here it has been assumed that b and x are not contained in the set of variables to be freed.

[2] Again, it has been assumed that x is no variable to be freed.

the model: if the user is sure to perform "good" abstractions, then it is recommended to use fine-grained abstractions, since the reduction of the model's complexity is best in comparison to other types. Otherwise, the user should select the middle-grained abstraction type, preserving more readability than the fine-grained abstraction type.

Independent of the degree of abstraction granularity, an iterative approach is suggested below.

1. extract cone of influence $\mathcal{K}^{\mathcal{S}}_{\mathcal{M}}$ of model $\mathcal{M}$ w.r.t. requirement $\mathcal{S}$. This can be done automatically. Let $\mathcal{V}^{\mathcal{K}^{\mathcal{S}}_{\mathcal{M}}}$ be the set of variables remaining in the model with respect to $\mathcal{K}^{\mathcal{S}}_{\mathcal{M}}$.
2. select the set $\mathcal{A} \subseteq \mathcal{V}^{\mathcal{K}^{\mathcal{S}}_{\mathcal{M}}}$ of variables to be freed
3. generate abstract model $\mathcal{M}^{\mathcal{A}}$ out of $\mathcal{M}$ automatically by supplying set $\mathcal{A}$.
4. try to verify specification $\mathcal{S}$ against $\mathcal{M}^{\mathcal{A}}$:
   (a) if model-checker concludes that $\mathcal{M}^{\mathcal{A}}$ fulfills $\mathcal{S}$, then also $\mathcal{M}$ fulfills $\mathcal{S}$.
   (b) if model-checker returns that $\mathcal{M}^{\mathcal{A}}$ does not fulfill $\mathcal{S}$, then examine error-path. If $\mathcal{M}^{\mathcal{A}}$ shows traces which are impossible in the concrete model $\mathcal{M}$ then remove the elements causing this unexpected behaviour from set $\mathcal{A}$ and return to (2) or (3), respectively.
   If the counterexample shows also a trace for $\mathcal{M}$, then either there is an error in the model or the specification is too strong.
   (c) if the model checker is not able to produce any output due to complexity, then put some new elements $a \in \mathcal{V}^{\mathcal{K}^{\mathcal{S}}_{\mathcal{M}}}$ in $\mathcal{A}$, "set" $\mathcal{M} := \mathcal{M}^{\mathcal{A}}$ and return to (1). By "setting" $\mathcal{M}$ to $\mathcal{M}^{\mathcal{A}}$ the cone of influence gets smaller, since due to the abstraction, some behavioural parts of the model—and hence some variables—will not be able to influence outputs anymore.

4.2 Example Abstraction and Verification

This subsection explains the application methodology of the abstraction techniques described above by example. We show, that the proof carried out using system verification can also be done using abstraction techniques, in fact on a 12-store SMS configuration.

To see which variables are not required to have their concrete values and hence can be freed, it is helpful to have a look into NORMAL_OPERATIONS, responsible for a "normal" run of the system. NORMAL_OPERATIONS is depicted in Figure 11.

The decision to release a store is taken in the submodule PERFORM_RELEASE_E. This decision is made with respect to a release order (Signal PAK_SEQUENCE), in which the release sequence is determined in a manner maintaining stability of the aircraft. This sequence (calculated in activity DEFINE_SEQUENCE_E) depends on the current balance (signal CURRENT_BALANCE) of the aircraft and a so-called package (Signal SELECTED_PACKAGE), in which the configuration of stores is defined in more

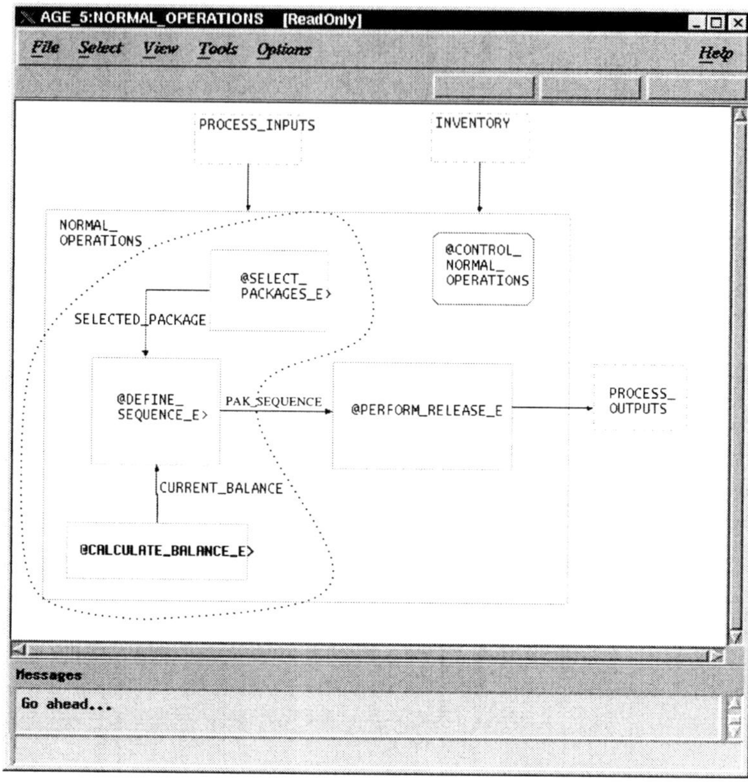

Fig. 11. Subactivity NORMAL_OPERATIONS

detail [CS97].

Consider again the specification in Figure 3. It is obvious that a release is influenced by the release sequence, but the *concrete* order is not required for verification, since the property must hold for *any* sequence. The same conclusion can be derived easily for the current balance of the aircraft or the number of stores to release in sequence. All variables representing these attributes can be freed. After performing this abstraction, verification successfully finished for the complete SMS 5-stores version after only 17 seconds. Also the 12-stores version fulfilled the property—the proof took 960 seconds. Note that no iteration of the scheme depicted on page 20 was required.

Table 1 shows the verification results with respect to different degrees of granularity. The middle column shows the number of state and input bits after performing abstraction and calculation of the cone of influence. The right-hand side shows the time required for verification in seconds. The last row shows the verification results for the 12-stores version of SMS. The attempt to verify the specification using coarser abstraction types failed due to complexity.

granularity	*state/input bits*	*time for verification (sec)*
middle	95/98	74.73
fine	89/102	16.72
fine (12)	157/268	960

Table 1. Verification results on abstract model

5 Conclusion

We have presented the STATEMATE verification environment and its application to the verification of an industrial application developed at British Aerospace. As of today, six other industrial applications in automotive as well as avionics have been verified—detecting errors in all cases, though designs had undergone substantial testing. We feel—for reasons elaborated in the introduction—that the complexity of ECU's makes it impossible for designers to come up with a complete set of attacks, and that as of today only model checking can provide 100% coverage wrt. a given set of safety-critical properties. Of course, this only shifts the completeness problem up one level—systematic ways of characterizing all safety requirements for a design need to be developed.

We currently see three reasons for introducing formal verification technology:

1. Model checking with the diagnostic capabilities discussed in the paper is a powerful debugging tool. Recalling the quantitative argument, one model-checker run covers all pathes of the model in one sweep, and counterexamples give clear indications where to spot the bug. By providing pre-defined templates for typical "debugging queries", the training effort to use model checking for debugging purposes can be reduced to a level, where every STATEMATE user can employ the tool—and cut down development costs.
2. Verification of safety properties was the key motivation for analysing capabilities of model-checking based verification technology. It is here, where the qualitative argument for model checking against testing comes into play: only the model-checker will find all possible attacks—a claim substantiated by the industrial evaluations cited above. Specifying safety-critical properties as symbolic timing diagrams does require training, an effort, which will be reduced in future releases of the toolset by providing specification-pattern libraries. For the application to complete ECU designs, the recently developed option of abstraction is important to reduce the number of system verification steps, which clearly requires more training than application of model checking, since local properties must be identified and specified by the designer.
3. Finally, a key use of the system verification technology rests in supporting IP-protection, a key factor in ECU developments distributed over manufacturer and supplier companies, as in avionics and automotive. Here, grey-box views of designs expressed as symbolic timing diagrams could be offered by suppliers as a means for manufacturers to check consistency with manufacturer-

defined requirements and a basis for integration with code designed by other sources, such as legacy code.

Currently, a number of extensions of the verification technology described in this paper are under development.

The verification engine will be enhanced to cope with a restricted range of data reasoning, such as required when analysing the controller's reaction on thresholds computed in a continuous submodel of the controller [BDG+98]. To this end, specification tools like MatrixX and Mathworks will be interfaced to SMI and SSL, hence providing access to the verification capabilities described in this paper. Other extensions deal with handling real-time characteristics of the target hardware.

We characterized formal verification technology as being on the edge of introduction to industrial design processes. Such a step is a major decision, which clearly must be subject to rigorous trials, substantiating confidence in the industrial relevance of the tool set. The capability of the STATEMATE verification environment described in this paper, in our opinion, does show the potential of introducing formal verification into the industrial design process.

References

[BBE98] Jürgen Bohn, Udo Brockmeyer, and Claus Essmann. SMI – system modelling interface. Technical Report CS95-31, Carl von Ossietzky Universität Oldenburg, 1998.

[BDG+98] J. Bohn, W. Damm, O. Grumberg, H. Hungar, and K. Laster. First-order CTL model checking. In *FSTTCS 98*, LNCS, 1998. To appear.

[Ben98] Albert Benveniste. Safety critical embedded systems design: the sacres approach. In *School on Formal Techniques in Real Time and Fault Tolerant Systems 1998: Material for the School.* Tech. Univ. of Denmark, 1998.

[BL69] J. R. Büchi and L. H. Landweber. Solving sequential conditions by finite-state strategies. *Trans. Amer. Math. Soc.*, 138:295–311, 1969.

[Bry92] Randal E. Bryant. Symbolic boolean manipulation with ordered binary-decision diagrams. *ACM Comp. Surveys*, 24:293–318, 1992.

[CES83] Edmund M. Clarke, E.A. Emerson, and A.P. Sistla. Automatic verification of finite state concurrent systems using temporal logic specifications: A practical approach. In *Procceedings of the 10th ACM Symposium on Principles of Programming Languages*, pages 117–126, 1983.

[CGL94] Edmund M. Clarke, Orna Grumberg, and David E. Long. Model checking and abstraction. In *ACM Transactions on Programming Languages and Systems*, volume 16, pages 1512–1542, September 1994.

[CS97] Geoffrey Clements and Eric Sefton. Deliverable report E3.A, BAe application specification. Technical report, EP 20897 Safety Critical Embedded Systems: From Requirement to System Architecture (SACRES), September 1997.

[DJHP98] Werner Damm, Bernhard Josko, Hardi Hungar, and Amir Pnueli. A compositional real-time semantics of STATEMATE designs. In W.-P. de Roever, editor, *Proceedings, International Symposium on Compositionality – The*

Significant Difference, Lecture Notes in Computer Science. Springer-Verlag, 1998.

[DJS95] W. Damm, B. Josko, and R. Schlör. Specification and verification of VHDL-based system-level hardware designs. In E. Börger, editor, *Specification and Validation Methods*, pages 331–410. Oxford Univ. Press, 1995.

[Döh97] Gert Döhmen. Deliverable report I1.4.A, The system specification language within SACRES. Technical report, EP 20897 Safety Critical Embedded Systems: From Requirement to System Architecture (SACRES), September 1997.

[Ess98] Claus Essmann. Optimierungen für smi-programme. Technical report, Carl von Ossietzky Universität Oldenburg, February 1998.

[FJ97] Konrad Feyeraband and Bernhard Josko. A visual formalism for real time requirement specifications. In *Proceedings of the 4th International AMAST Workshop on Real-Time Systems and Concurrentand Distributed Software, ARTS'97, Lecture Notes in Computer Science 1231*, pages 156–168, 1997.

[HGD95] H. Hungar, O. Grumberg, and W. Damm. What if model checking must be truly symbolic. In P. Camurati and H. Eveking, editors, *CHARME 95*, LNCS 987, pages 1–20. Springer, 1995.

[HN95] D. Harel and A. Naamad. The STATEMATE semantics of statecharts. Technical Report CS95-31, The Weizmann Institute of Science, Rehovot, 1995.

[Hol98] H. Holberg. OFFIS verification tool-set (An installation tutorial). Technical report, Kuratorium OFFIS e.V., Oldenburg, Germany, 1998.

[Jos93] Bernhard Josko. Modular specification and verification of reactive systems. Carl von Ossietzky Universität Oldenburg, 1993. Habiltationsschrift.

[Jos97] B. Josko. Symbolic timing diagram specification: A tutorial. Technical report, Kuratorium OFFIS e.V., Oldenburg, Germany, 1997.

[LSU95] Roger Lipsett, Carl Schaefer, and Cary Ussery. *VHDL: Hardware Description and Design*. Kluwer Academic Publishers, 1995.

[McM93] Kenneth L. McMillan. *Symbolic Model Checking*. Kluwer Academic Publishers, 1993.

[PSS98] A. Pnueli, M. Siegel, and O. Shtrichman. Translation validation for synchronous languages. In *ICALP 98*, LNCS 1443. Springer, 1998.

[Win98] K. Winkelmann. Deliverable report I2.B, The SACRES design methodology for safety-critical systems. Technical report, EP 20897 Safety Critical Embedded Systems: From Requirement to System Architecture (SACRES), 1998.

[Wol86] Pierre Wolper. Expressing interesting properties of programs in propositional temporal logic. In *Proceedings of the 13th Annual ACM Symposium in Principles of Programming Languages*, 1986.

Safety and the Millennium Bug

Tony Foord and Harry Smith
4-sight Consulting

Abstract

This paper describes practical examples of year 2000 induced failures (some of which have already occurred), arising from calculating with and manipulating dates. We also describe some of our experiences in dealing with the fact that errors in dates do cause problems within safety-related systems, and that the consequences arising from such problems are of an unpredictable nature.

1. Introduction

"Software does not depend on dates to function - ... This whole band-wagon has been a windfall for the software consulting industry that knows a gift horse when it sees one." wrote Dr Mark Scibor-Rylski [Scibor-Rylski 1998]. It is tempting to agree with this, when there is so much hype in the media about death and destruction over the Millennium midnight, being caused by everything from commercial aeroplanes to domestic washing machines. Unfortunately there are numerous examples of deaths caused by flaws in software, though nothing like the number of deaths from other causes [Leveson 1995]. Equally there are examples where the software is known to have bugs, e.g. ignore leap year transitions, but where the overall functionality of the system is not compromised.

So, how could errors with dates and / or time cause problems with safety-related systems? What evidence is there, and what should we do?

This paper describes practical examples of year 2000 induced failures (some of which have already occurred), arising from calculating with and manipulating dates. We also describe some of our experiences in dealing with the fact that errors in dates do cause problems within safety-related systems, and that the consequences arising from such problems are of an unpredictable nature.

2. What is the Problem?

There are numerous, well-documented, examples of Millennium bugs in IT systems, however this paper is concerned with what are frequently, and unhelpfully, called "embedded systems". By these we mean real-time, control, monitoring or protection systems, or what are sometimes called Programmable Electronic Systems (PES) [IEC 1998]. We include in these "decision support" systems since these are often an important part of the monitoring process, and may strongly influence the actions of human operators.

2.1. Century roll-over / leap year.

Within programmable systems, all dates and times are defined as offsets from an agreed datum time (e.g. 1 A.D.). For example:-

- the NPL MSF Time Service at Rugby transmits a signal with Binary-Coded-Decimal Year (00-99), Month (01-12), Day-of-Month (01-31) and Day-of-Week (0=Sunday to 6=Saturday) as well as the time in hours, minutes and second [Laverty 1997].
- Microsoft applications packages often take 1970 as the default start year to which is added an accumulating offset.
- Unix, C and C+ all use a date base of 01/01/1970 with an accumulated second counter acting as the offset (this 32 bit second counter will overflow at 03:14:08 on Tuesday 19 January 2038).

Five centuries ago astronomers calculated the numbers of days in the year to an accuracy that we still find impressive, even though we now know that there is some variation from year to year in the time it takes for one revolution of the earth around the sun. The 15th century astronomers worked out that to maintain, over many centuries, the first of January at mid-winter in Europe we needed to add 97 days every 400 years to a standard year of 365 days. Hence the pattern of a leap year every 4 years except 1700, 1800, 1900, 2100 etc.

A more formal definition of both the leap year and "Year 2000 Conformity" was first detailed in [BSI 1997-1] (Appendix 1) which has since been revised with document [BSI 1998-1] (Appendix 2).

Unfortunately, even though the pattern of leap years is agreed, there are conflicting conventions about how a date is written down; North Americans use mm/dd/yy and Europeans use dd/mm/yy. When a date is used in a PES the methods vary much more widely e.g. the standards for COBOL, Bar-codes, BACS, credit cards, etc. all insist that the year field has only 2 digits. Other standards include storing a date as the year plus the number of days (YYDDD); or counting the days, minutes, seconds or some fraction of a second from a particular datum (e.g. 1/1/1900). All PES have limits on the size of integers that can be stored, displayed and printed and thus there will be limits on the manner in which dates can be manipulated.

We and many others have discovered problems of:-

- Ambiguity - when a two digit date is used in a calculation involving duration (e.g. trend recording or metering) then subtraction of any other two digit year from 00 will give a negative duration and erroneous results;
- Discontinuity - when crossing a day transition at midnight or during changes between summer and winter time [Kletz 1995] unexpected behaviour can occur in the control, monitoring and protection system, even though dates may not be directly involved in any of the calculations. These problems are usually associated with dates the system cannot handle because they are either missing (e.g. 29 February) or are beyond the range

of the system (overflow) or time appears to be reversed (e.g. the end of summer time or restoring today's date following tests using future dates).

2.2. Other dates.

Despite publicity emphasising the year 2000 roll-over and its associated leap year status there are many other problematic dates (Appendix 3) not all of which relate directly to year 2000 but which can have similar consequences through common cause failure. An example of the latter is with the much implemented Geographical Positioning System (GPS) which will exhibit roll-over problems when its offset counter overflows during the transition between 21 August 1999 and 22 August 1999.

Users and operators are very adaptable and will use systems for unplanned purposes. Thus the financial services industry have found examples of dates which are used as signals or switches in IT systems e.g. "1/1/00" in a "date of termination" field to indicate that the contract does not have a termination date. Similar problems have been found in IT systems for "9/9/99", "31/12/99" and other dates selected by users. We have not yet found this type of date usage in control, monitoring or protection systems.

2.3. Unpredictability.

Problems of unpredictability in software are not new. "Arbitrary limits", ambiguities caused by "weak typing" and "exception handling" all produce results unexpected by the user or operator. In fact, these results are entirely predictable, as the software will reproduce them consistently, but from the operator's point of view they do not match the operator's mental model of how the system should behave, and may be very different from the behaviour intended by the designer. However failures in year 2000 are unique in that the common cause failure of a system will not necessarily be repeatable and the consequential interactions with both related and apparently non-related systems will provide considerable confusion and the potential for dangerous incidents to occur.

2.4. Possible Consequences of Non-conformity.

The consequences of the unintended behaviour may be much more serious that simply printing "1900" instead of "2000":-

- A protection system may not respond to a demand to operate (e.g. shutdown)
- A control system may not maintain a process within the desired limits and move into an unsafe region of operation.
- A monitoring system may produce mis-leading information which causes the operators to take the wrong action, or as the case of the smelter example below, to decide to shutdown because they were no longer confident in the system used to monitor critical processes.
- A controlled process, dependent on 3rd party utilities, may fail to function through a failure in one or more of the supporting utilities.

As well as the application code PES use:-

- Data tables
- Operating system
- Run-time libraries (including I/O routines)
- Compilers
- Test tools

and safe operation may depend on the correctness of all of these.

2.5. Common cause failures.

A vital part of designing and maintaining any safety-related system is to avoid situations where a single failure (e.g. the supply of power) will cause simultaneous failure of both the control system and the protection system. It is this potential for a single common cause failure across many business providers, both related directly and indirectly, that gives the problem of year 2000 its unique importance.

2.6. Problems Internal to a System.

Much of the concern about the Millennium bug, is that virtually all PES include internal clocks which measure or count time from which to compute dates, and they will all transition to problem dates (e.g. 1/1/00, 29/2/00) at the same time. The difficulty is deciding if the dates are used by significant parts of the software; and, if used, are they used and interpreted correctly, and, if not, could the resulting non-conformity lead to a significant failure.

2.7. Problems External to a System (i.e. 3rd party support).

All safety-related systems rely on some degree of monitoring (automatic or manual) to ensure the health of the system. There can be problems if the monitoring or support process has Millennium bugs. These include problems with the supply chain for that process e.g. spares which are not available, because of problems of supply, or after a system has been upgraded to achieve year 2000 conformity, a non-compliant spare is installed during a future maintenance operation.

2.8. Redundancy.

Redundancy is a well established method of protecting against random hardware failures - but should both components of a 'redundant' control system contain Real Time Clocks (RTC) then the reliability argument will not hold true for any failure mechanism generated through a year 2000 failure. There is also ample evidence that multiple versions of software provide little additional protection [Leveson 1995] and also all versions of a particular software package may well use the same date functions.

2.9. Diversity.

Another well established design technique is to make the redundant system completely different e.g. one electronic and the other hydraulic. Both are then very unlikely to fail from a common cause. This works particularly well if the redundant system contains no software. However, if both diverse parts contain micro-processors (now normal practice in hydraulic systems) there may be less diversity than intended because the RTCs within the diverse systems will roll-over together causing a common cause failure of the entire system.

2.10. Concurrent failures of a number of related or un-related systems.

Thus our concern is to avoid simultaneous failures of a number of systems, even if these systems are apparently unrelated. Many accidents are the result of complex circumstances involving multiple mistakes or un-anticipated events that combine to produce a disaster. Microprocessors are now so widely used that dealing with potential concurrent mis-behaviour is a very serious issue.

3. Industrial action and methods of tackling the problem.

The potential problem arising from the year 2000 transition was widely recognised in 1995, and in some cases much earlier than that. Those industries which include hazardous operations or process hazardous materials have all used semi-formal methods to quantify the problem and the same methods can be successfully to decide on the business risk from any system failing through year 2000. The three part method described in [HSE 1998-1] is:

1. Action and Analysis (which includes Risk Assessment)
2. Strategy and Planning
3. Implementation and Test

Action 2000's seven step method is:

1. Understanding
2. Prioritise (including Risk Assessment)
3. Assess
4. Plan
5. Implement
6. Test
7. Install

All the industrial methods are similar and involve making an inventory; classifying the inventory using a risk assessment process; liasing with suppliers to assess non-conformity; deciding to live with the problem or apply a work-around (mitigation), fix the problems, replace or retire the systems; implementing the resulting plan; testing all critical systems (both those changed and those apparently compliant); installing any fixes or upgrades; and finally, and most importantly, a contingency plan to cover unforeseen failures. It is impossible to identify all PES having an influence on any particular system and which might well cause the system to mal-function. No-one should entertain the belief that everything will have been fixed.

3.1. Risk assessment.

The biggest difficulty is the sheer scale of the problem. Checking the billions of chips installed world-wide is not only uneconomic, it is unachievable. Thus, all the methods used to combat potential year 2000 failures include a process of risk assessment e.g. [HSE 1998-1]

1. Assessing Criticality - the impact or consequences of failure
2. Assessing Functionality - does the PES include date and time
3. Assessing Construction - what is the likelihood or probability

The impact and likelihood can be combined by multiplication to give a risk classification of each item of the inventory. Even though the process of risk assessment may be based on detailed checklists, or well defined criteria, it is still subjective. Thus a simple table may be used instead of numerical calculation to sort the inventory into risk categories A, B, C, or D e.g.

Risk Category	**A, B, C, or D**		**Impact**	
		Low	Medium	High
Likelihood	High	C	B	A
based on PES	Medium	C	B	A
construction	Low	D	C	B

[HSE 1998-1] uses seven risk categories 0 to 6, but then groups them into the same 4 risk categories as the table above i.e. High (6); Upper medium (4-5); Lower medium (2-3); and Low (0-1) because different actions are recommended for only 4 categories:

Risk	**Acceptance Criteria**	**Explanation**
A	Complies with BSI DISC PD2000-1:1998 Uses agreed date representation, including any agreed date window Does not fail unsafe Meets all safety requirements	These PES have been identified as high risk. Workarounds are not acceptable.
B	Uses agreed date representation, including any agreed date window Does not fail unsafe Meets all safety requirements	Evidence exists to ensure safe operation even if the system is not fully compliant, e.g a workaround is available
C	Need not use agreed date representation Contingency plans exist for failures Meets all safety requirements	Safety can be assured by alternative means
D	Failure does not impact safety	No direct impact on safety

3.2. Interactions with systems over which you have no control.

There has been considerable debate over the reliability of the national infrastructure and the major utilities. While the utility companies have never guaranteed an uninterrupted supply, they recognise their importance to modern living. Nearly all of them have grouped together to co-operate to solve year 2000 problems. The electricity, gas, telecommunications and water utilities have all joined the Utilities Group of the UK Year 2000 Interest Group and have now been meeting regularly since 1997. [HSE 1998-1] states on page 14: " From the evidence at present (1997), we should conclude that the major utility providers will continue to successfully provide services through the Year 2000 change-over, but local facilities will need to be re-examined, such as local telephone networks, back-up generators, boiler plant, ventilation systems and so on." In other words we have focused on the local facilities over which we can exercise some degree of control.

3.3. The need for and limit to testing.

We should not rely on the common assumption that the solution is to test everything. There is ample evidence that testing is only one method (and not the most effective method) of finding errors in software [Leveson 1995]. In our investigations we have used code inspection, verification, and validation including full and simulated testing.
Testing is particularly expensive because year 2000 tests âre difficult to automate. Tools are available to run test scripts, but not to reboot or re-IPL a PES for different system dates.

As a point of warning to the unwary there are numerous examples of people mistakenly testing the production system without having first evaluated potential consequences. When the production system is restored to today's date there are then problems, as the designer did not cater for time travelling in reverse or the licence agreement precludes such an action. A separate test and development system or engineering workstation for testing is strongly recommended. Very good configuration control and change management is needed to ensure:

- the test and production systems match
- the production system is not changed and made non-compliant.

3.4. Contingency planning.

Industries operating in environments likely to be hazardous and which implement safety-related and safety-critical systems will have existing procedures for handling disasters. Documents such as:

- Safety Case - a legal requirement
- Emergency situation plan - to prevent an failure prior to it achieving a magnitude to become disaster
- Disaster action plan - The ultimate to minimise damage to Safety, Health and Environment (SHE) from failure to contain an emergency situation.

These documents will already be in place and will require to be reviewed and updated in light of the outcomes from the risk assessment of systems likely to malfunction or cease to function during a date transition. It is vital that those responsible for the ownership of the systems at the transition dates should be fully conversant and in agreement with the upgraded documents.

3.5. Embedded systems versus IT systems.

Whereas IT systems are generally well defined and are 'obviously' programmable; embedded PES within the process and automation control industries are not so easy to identify. In many instances there is no screen or keyboard to provide the hint and, as most 'systems' are produced from many sub-systems, with each and every one produced and supplied by one or more subcontractors, identification of an embedded microprocessor becomes a difficult and time consuming process which carries no guarantee of 100% success.

4. How to find them.

We have used all the following methods to find installed micro-processors.

4.1.1 Detailed discussion with vendor and their suppliers; information from this source needs very delicate handling as, if a vendor has bought in electronic hardware to integrate with their own systems they may not be aware of what resides in the bought-in 'black box' and may respond incorrectly to any questioning. Indeed it can be impossible to find a person within a vendor's organisation with any knowledge of the intricate detail of the systems they have provided.

4.1.2 Investigation of available documentation; technical drawings and parts-lists can further narrow down the investigation. However there must be a high confidence as to the accuracy of such documentation. It is normally the rule that the longer the time period since commissioning the poorer the status of documentation.

4.1.3 Physical investigation of actual hardware can be very useful provided that the investigator knows where to look and that the application is accessible. Access is highly probable on land based installations but, in many instances, is impossible with off-shore installations.

4.1.4 discussion with service/commissioning engineers; For those year 2000 searchers fortunate to be investigating a green field installation, access to commissioning engineers can be the most successful method of all.

4.2. How to identify clock functionality.

- For PES to maintain a clock function throughout periods of power failure there is the need for the designer to provide a back-up power source, namely a battery or large capacitor. Identification of a battery or large capacitor is evidence of the need to investigate further and ascertain whether the clock functionality is restricted to clock synchronisation pulse generation or whether such clock 'ticks' are used to build up time and date parameters for manipulation within a software or firmware program. Examination of hardware and software functional specifications along with 'walkthroughs' of the actual software program listings (taking care that they themselves are up-to-date) will be required.

- With some control packages information will be claimed by the vendor to be their 'intellectual property' and a decision will be required as to the comprehensiveness of the vendor's compliance statement. This applies very much to proprietary operating systems.

4.3. The need for mitigation or workarounds

Having identified a PES and determined its potential to cause harm then a decision has to be taken as to the way forward. It is unacceptable to retain the status quo. A decision may be :

- Do nothing. However responsibility for the subsequent consequences arising from a year 2000 failure will remain with you as insurance is unlikely to be available. It is good practice to document the reasoning behind the decision to do nothing.
- If a system is known to be non-compliant, and after testing it has been shown to have no effect on system functionality or to the functionality of other systems to which it communicates then no further action, other than to document to test results, may be deemed necessary .
- It is known to be non-compliant and does have an effect on functionality so:
    1. Provide a design change and then re-test.
    2. Totally replace the equipment with a compliant form.

5. Third party vendor assessment.

No business can exist on its own and, inevitably, will require support from a third party over which it can exercise minimal influence. (e.g. provision of electrical power from a service provider). All such vendors will require some form of auditing to gain assurance that they are managing their both own business risk from probable year 2000 failures, and also that of their own suppliers, and that no failure of the third party's business could compromise the primary system.

6. Examples.

6.1. Where safety could be influenced directly by date or time malfunction.

Summer time changes

A computer was controlling a batch reaction on a chemical plant during the night when summer time ended and the clocks had to be put back an hour. The operator reset the clock in the computer so that the time indicated 2 am instead of 3 am.

The computer then shut the plant down. One hour later the computer restarted the plant without warning [Kletz 1995].

Leap year

On 31 December 1996 the computer control system on a New Zealand smelter started displaying inconsistent data and the operators lost confidence that they had control of the smelter and shut it down. Some time later the sister plant in Tasmania experienced the same problems and the operators made the same decision. Subsequent investigation revealed that the bespoke application code was programmed for years of 365 days and could not handle the 366th day.

Crane load monitoring

Some cranes record in their PES the load and distance travelled for each crane movement. No date or time stamp is used, so there is no date related functionality and year 2000 conformity was not seen as an issue. When the memory is full, further use of the crane is barred until the data is down-loaded to a portable data-logger, and the memory reset. The data-logger is not year 2000 compliant and, if uncorrected, will impact the operation of the crane.

Hull-stress analysis and control of cargo and ballast.

Within an off-shore oil production and storage vessel there is a need to ensure that the on-board ballast and extracted oil is distributed correctly throughout the various storage tanks to ensure that the stresses within the hull do not exceed design levels. The consequences of uneven distribution could cause failure of the hull's integrity, leading to loss of ballast and/or crude oil into a sensitive environment and, in the extreme, could lead to total vessel loss. The control system had been integrated from equipment sourced from three different vendors consisting of:-

- tank level control,
- cargo and ballast pumping,
- stress analysis algorithm.

Analysis of the three systems indicated that no RTC existed within the cargo and ballast pumping controls but a RTC was present within the other two systems and both were tested. Both the stress analysis and the tank level control exhibited a date failure on the century roll-over. The non-compliance within the stress analysis involved both the hardware (BIOS) and software necessitating its changeout to compliant hardware and upgraded software. The non-compliance within the level control proved to be cosmetic in that the only function affected was the MMI (on-screen date). Mitigation in this case was to ensure an operator will be in attendance at problematic dates to correctly re-enter the screen date.

Power generation off-shore.

The main power plant on an off-shore platform consists of a generator driven by a gas turbine engine. The control functionality is distributed over three microprocessors, two of which provide all the control functionality whilst the third provides the necessary communication with the platform's main SCADA system. Inter microprocessor communication is via data highways. The MMI is provided by a proprietary application package configured for the specific application. This system, classified as essential for the safety and emergency evacuation of the platform, underwent independent witness testing of its capability to handle all

known problematic dates. All microprocessors and the MMI failed. The latter extensively whilst the former simply rejected all requests to enter 29 February.
It proved necessary to change out the MMI to one with proven compliance. Testing of the microprocessors showed that, despite the failure to enter correct leap day, all other functionality was executed correctly and no hardware/firmware upgrade was deemed necessary.

6.2. Where unsafe practice has been identified as a by-product of carrying out Y2K inventory.

The work required to identify year 2000 issues does have an appreciable beneficial spin-off in that the close attention to detail required to build up an inventory of PES brings all areas of a process under scrutiny and provides an opportunity to identify and rectify other safety issues.

Moisture analysis of dry gas stream.

During the extraction of natural gas and its subsequent drying prior to re-injection into the well (achieving improved oil extraction on difficult reservoirs) it is essential to maintain the moisture content at a low level. Failure to do so will lead to failure of gas compressor seals and a possible very large gas release to the atmosphere that will endanger personnel and the environment. Y2K investigations identified that there was only one moisture analyser present leading to the possibility of a single mode failure. The installation was upgraded to a dual redundant measurement.

O_2 analysis of inert blanket gas generation.

Crude oil is stored in tanks where the void above the oil level is filled with inert gas, generally diesel fumes. A measurement of the O_2 content of this inert gas is required to ensure that it never exceeds the level where an explosion occurs leading to the destruction of the tank and its surrounding environs. In this particular example two diesel boilers were utilised with an O_2 analyser in each train. Due to the installation construction when only one boiler was in use the O_2 measurement was demoted from a dual redundant measurement to a single point measurement. Due to the potential for single mode failure of the safety circuit the installation was modified.

7. Conclusions.

The Millennium bug can affect safety-related systems. In many cases the problem is not significant and in other cases the problem may be serious, but can be corrected by a simple workaround. In a very few cases the problem is both serious and cannot be corrected by a simple workaround. A software fix or upgrade is essential to maintain the integrity of the safety-related system. The problem is not restricted to the Millennium midnight. Problems have been identified with other dates, and at least those listed in Appendix 3 should be checked.

Although checking is time consuming, the number of safety-related systems on most installations is small and can be tackled in a few months. The management challenge is to find the necessary resources, and to focus on safety.

8. References.

[BSI 1997-1] *A Definition of Year 2000 Conformity Requirements*. BSI - DISC PD2000-1, UK, 1997.

[BSI 1998-1] *A Definition of Year 2000 Conformity Requirements*. BSI - DISC PD2000-1:1998, UK.

[HSE 1998-1] *Safety and the year 2000*. HSE, ISBN 0 7176 1491 3, UK, 1998.

[IEC 1998] *Functional safety: safety-related systems*. ISO/IEC DIS 61508, 1998

[Kletz 1995] Kletz TA: *Computer Control and Human Error*. IChemE, UK, 1995, page 20.

[Laverty 1997] Laverty J: *Millennium Compliance*. Newsletter from the National Physical Laboratory, Centre for Electromagnetic and Time Metrology, Time and Frequency Services, UK, 13 January 1997.

[Leveson 1995] Leveson N: *Safeware: system safety and computers*. Addison-Wesley Publishing Company, Inc, USA, 1995.

[Scibor-Rylski 1998] Scibor-Rylski M: *The emperor's new software*. IEE Review, UK, 17 September 1998, page 219.

9. Further reading

[BSI 1997-2] *Managing year 2000 conformity*. BSI - DISC PD2000-2:1997, UK.

[BSI 1998-2] *Understanding the century date-change problem*. BSI - DISC PD2000-3:1998, UK.

[HSE 1998-2] *Health and safety and the year 2000 problem, Guidance on year 2000 issues as they affect safety-related control systems*. HSE, INDG267, ISBN 0 7176 1570 7, UK, 1998.

[HSE 1998-3] *Testing safety-related control systems for year 2000 compliance*. HSE, ISBN 0 7176 1596 0, UK, 1998.

[IEE 1997] *Embedded systems and the Year 2000 Problem*. IEE Report, ISBN 0 85296 930 9, UK, 1997.

Appendix 1 - DISC PD2000-1

A Definition of Year 2000 Conformity Requirements

A DEFINITION OF YEAR 2000 CONFORMITY REQUIREMENTS

Introduction

This document addresses what is commonly known as Year 2000 conformity (also sometimes known as century or millennium compliance). It provides a definition of this expression and requirements that must be satisfied in equipment and products which use dates and times.

It has been prepared by British Standards Institution committee BDD/1/-/3 in response to demand from UK industry, commerce and the public sector. It is the result of work from the following bodies whose contributions are gratefully acknowledged: BT, Cap Gemini, CCTA, Coopers & Lybrand, Halberstam Elias, ICL, National Health Service, National Westminster Bank.

BSI-DISC would also like to thank the following organizations for their support and encouragement in the development of this definition: taskforce 2000, Barclays Bank, British Airways, Cambridgeshire County Council, Computer Software Services Association, Department of Health, Ernst & Young, Federation of Small Businesses, IBM, Id, National Power, Paymaster Agency, Prudential Assurance, Reuters, Tesco Stores.

While every care has been taken in developing this document, the contributing organizations accept no liability for any loss or damage caused, arising directly or indirectly, in connection with reliance on its contents except to the extent that such liability may not be excluded at law. Independent legal advice should be sought by any person or organization intending to enter into a contractual commitment relating to Year 2000 conformity requirements.

THE DEFINITION

Year 2000 conformity shall mean that neither performance nor functionality is affected by dates prior to, during and after the year 2000.

In particular:

Rule 1. No value for current date will cause any interruption in operation.

Rule 2. Date-based functionality must behave consistently for dates prior to, during and after year 2000.

Rule 3. In all interfaces and data storage, the century in any date must be specified either explicitly or by unambiguous algorithms or inferencing rules.

Rule 4. Year 2000 must be recognized as a leap year.

AMPLIFICATION OF THE DEFINITION AND RULES

General Explanation

Problems can arise from some means of representing dates in computer equipment and products and from date-logic embedded in purchased goods or services, as the year 2000 approaches and during and after that year. As a result, equipment or products, including embedded control logic, may fail completely, malfunction or cause data to be corrupted.

To avoid such problems, organizations must check, and modify if necessary, internally produced equipment and products and similarly check externally supplied equipment and products with their suppliers. The purpose of this document is to allow such checks to be made on a basis of common understanding.

Where checks are made with external suppliers, care should be taken to distinguish between claims of conformity and the ability to demonstrate conformity.

Rule 1

1.1 This rule is sometimes known as *general integrity*.

1.2 If this requirement is satisfied, roll-over between all significant time demarcations (e.g. days, months, years, centuries) will be performed correctly.

1.3 *Current date* means today's date as known to the equipment or product.

Rule 2

2.1 This rule is sometimes known as *date integrity*.

2.2 This rule means that all equipment and products must calculate, manipulate and represent dates correctly for the purposes for which they were intended.

2.3 The meaning of *functionality* includes both processes and the results of those processes.

2.4 If desired, a reference point for date values and calculations may be added by organisations; e.g. as defined by the Gregorian calendar.

2.5 No equipment or product shall use particular date values for special meanings: e.g. "99" to signify "no end value" or "end of file" or "00" to mean "not applicable" or "beginning of file".

Rule 3

3.1 This rule is sometimes known as *explicit/implicit century*.

3.2 It covers two general approaches:

(a) explicit representation of the year in dates: e.g. by using four digits or by including a century indicator. In this case, a reference may be inserted (e.g. 4-digit years as allowed by ISO standard 8601:1988) and it may be necessary to allow for exceptions where domain-specific standards (e.g. standards relating to Electronic Data Interchange, Automatic Teller Machines or Bankers Automated Clearing Services) should have precedence.

(b) the use of inferencing rules: e.g. two-digit years with a value greater than 50 imply 19xx, those with a value equal to or less than 50 imply 20xx. Rules for century inferencing as a whole must apply to all contexts in which the date is used, although different inferencing rules may apply to different date sets.

General Notes
For Rules 1 and 2 in particular, organisations may wish to specify allowable ranges for values of current date and dates to be manipulated. The ranges may relate to one or more of the feasible life-span of equipment or products or the span of dates required to be represented by the organisation's business processes. Tests for specifically critical dates may also be added (e.g. for leap years, end of year, etc). Organisations may wish to append additional material in support of local requirements.

Where the term century is used, clear distinction should be made between the "value" denoting the century (e.g. 20th) and its representation in dates (e.g. 19xx); similarly, 21st and 20xx.

DISC is a part of the British Standards Institution
BSI, 389 Chiswick High Road, London W4 4AL
Tel: 0181 996 9000

Appendix 2 - **DISC PD2000-1:1998**
A Definition of Year 2000 Conformity Requirements

A DEFINITION OF YEAR 2000 CONFORMITY REQUIREMENTS

Preamble to the Summer 1998 amendment
BSI DISC originally published PD2000-1 in January 1997 and it has been widely adopted. A review of the document was conducted by the responsible committee (BDD/1/3) in the spring of 1998 taking into account comments received. The committee considered that amendments to the fundamental conformity requirements were neither necessary nor desirable. The Definition and the four Rules are unchanged but, to add value to the document and aid its interpretation, the Amplification sections have been amended. This document, PD2000-1:1998, replaces the previous version of PD2000-1 but does not change its requirements. An additional document PD2000-4, entitled "PD2000-1 in Action" will provide further information on PD2000-1:1998 together with information on its use.
Paragraph numbers have been enhanced in the Amplification section to aid referencing and substantial revisions to the document are indicated by side lines against the changed text.

Introduction
This document addresses what is commonly known as Year 2000 conformity (also sometimes known as century or millennium compliance). It provides a definition of this expression and requirements that must be satisfied in equipment and products which use dates and times.
It has been prepared by British Standards Institution committee BDD/1/3 in response to demand from UK industry, commerce and the public sector. It is the result of work from the following bodies whose contributions are gratefully acknowledged: BT, Cap Gemini, CCTA, PricewaterhouseCoopers, Halberstam Elias, ICL, National Health Service, National Westminster Bank. Additionally, BSI DISC acknowledges the support of the Electronics and Information Industries Forum (EIIF), Action 2000, Taskforce 2000 and Digital Equipment as well as the original bodies for their participation in the review of this document.
BSI DISC would also like to thank the following organizations for their support and encouragement in the development of this definition: Barclays Bank, British Airways, Cambridgeshire County Council, Computer Software Services Association, Department of Health, Ernst & Young, Federation of Small Businesses, IBM, ICI, National Power, Paymaster Agency, Prudential Assurance, Reuters, Tesco Stores.

While every care has been taken in developing this document, the contributing organizations accept no liability for any loss or damage caused, arising directly or indirectly, in connection with reliance on its contents except to the extent that such liability may not be excluded at law. Independent legal advice should be sought by any person or organization intending to enter into a contractual commitment relating to Year 2000 conformity requirements.

This entire document or the Definition section (including the four Rules) may be freely copied provided that the text is reproduced in full, the source acknowledged and the reference number of this document is quoted. It is recommended that the Amplification section be included. References to "PD2000-1:1998" shall be interpreted as meaning the entire document.

THE DEFINITION

Year 2000 conformity shall mean that neither performance nor functionality is affected by dates prior to, during and after the year 2000.

In particular:

Rule 1 No value for current date will cause any interruption in operation.

Rule 2 Date-based functionality must behave consistently for dates prior to, during and after year 2000.

Rule 3 In all interfaces and data storage, the century in any date must be specified either explicitly or by unambiguous algorithms or inferencing rules.

Rule 4 Year 2000 must be recognized as a leap year.

AMPLIFICATION OF THE DEFINITION AND RULES

1 General Explanation

1.1 Problems can arise from some means of representing dates in computer equipment and products and from date-logic embedded in purchased goods or services, as the year 2000 approaches and during and after that year. As a result, equipment or products, including embedded control logic, may fail completely, malfunction or cause data to be corrupted.

1.2 To avoid such problems, organizations must check, and modify if necessary, internally produced equipment and products and similarly check externally supplied equipment and products with their suppliers. The purpose of this document is to allow such checks to be made on a basis of common understanding.

1.3 Where checks are made with external suppliers, care should be taken to distinguish between claims of conformity and the ability to demonstrate conformity.

2 Amplification of the definition

2.1 PD2000-1 (all editions) is solely concerned with the performance and functionality of a single version, release or system. It does not address differences in performance or functionality between different versions, releases or systems.

2.2 Variations in performance immeasurably small in the context of use do not make a version, release or system non-conformant.

3 Amplification of the Rules

3.1 Rule 1

3.1.1 This rule is sometimes known as *general integrity.*

3.1.2 If this requirement is satisfied, roll-over between all significant time demarcations (e.g. days, months, years, centuries) will be performed correctly.

3.1.3 *Current date* means today's date as known to the equipment or product, i.e. the actual date of operation [NOTE - this refers to normal operation and does not prevent testing.]

3.2 Rule 2

3.2.1 This rule is sometimes known as *date integrity.*

3.2.2 This rule means that all equipment and products must calculate, manipulate and represent dates correctly for the purposes for which they were intended.

3.2.3 The meaning of *functionality* includes both processes and the results of those processes.

3.2.4 If desired, a reference point for date values and calculations may be added by organizations; e.g. as defined by the Gregorian calendar.

3.2.5 No equipment or product shall use particular date values for special meanings; e.g. "99" to signify "no end value" or "end of file" or "00" to mean "not applicable" or "beginning of file" unless the values in question lie outside its possible date range.

3.3 Rule 3

3.3.1 This rule is sometimes known as *explicit/implicit century.*

3.3.2 It covers two general approaches:

(a) explicit representation of the year in dates: e.g. by using four digits or by including a century indicator. In this case, a reference may be inserted (e.g. 4-digit years as allowed by ISO 8601:1988) and it may be necessary to allow for exceptions where domain-specific standards (e.g. standards relating to Electronic Data Interchange, Automatic Teller Machines or Bankers Automated Clearing Services) should have precedence.

(b) **the use of inferencing rules**: e.g. two-digit years with a value greater than 50 imply 19xx, those with a value equal to or less than 50 imply 20xx. Rules for century inferencing as a whole must apply to all contexts in which the date is used, although different inferencing rules may apply to different date sets. Where any date element is represented without a century, the correct century shall be unambiguous for all manipulations involving that element.

3.4 Rule 4

3.4.1 A leap year is defined in ISO 8601:1988 (amended in 1991) as follows: "year, leap: In the Gregorian calendar, a year which has 366 days. A leap year is a year whose number is divisible by four an integral number of times, except that if it is a centennial year it shall be divisible by four hundred an integral number of times."

3.4.2 Thus, for example, 2000 is a leap year but 1900 is not.

4 General Notes

4.1 For Rules 1 and 2 in particular, it is recommended that the allowable ranges for values of current date and dates to be manipulated be documented, recognizing that all systems have some limitation on the valid date ranges. The ranges may relate to one or more of the feasible life-spans of equipment or products or the span of dates required to be represented by the organization's business processes.

4.2 Tests for specifically critical dates may also be added (e.g. for leap years, end of year, etc.). Organizations may wish to append additional material in support of local requirements.

4.2 Where the term "century" is used, clear distinction should be made between the "value"denoting the century (e.g. 20th) and its representation in dates (e.g. 19xx); similarly, 21st and 20xx.

DISC is a part of the British Standards Institution ISBN 0 580 305 074
BSI, 389 Chiswick High Road, London W4 4AL
Tel: 0181 996 9000

Appendix 3 - Problematic dates

31 December 1998 roll over to 1 January 1999 - possible end of file character
21 August 1999 roll over to 22 August 1999 - GPS time accumulator overflow
31 December 1999 roll over to 1 January 2000 - new century
28 February 2000 roll over to 29 February 2000 - leap day
29 February 2000 roll over to 1 March 2000 - first day after leap day
31 December 2000 roll over 1 January 2001 - end of first year
28 February 2001 roll over 1 March 2001 - first non leap day of new century
28 February 2004 roll over to 29 February 2004 - next leap year day
29 February 2004 roll over to 1 March 2004 - day following second leap day of new century
18 January 2038 roll over to 19 January 2038 - overflow of Unix, C and C+ systems.

For other futuristic dates see the IEE web site at http://www.iee.org.uk/2000risk/

Integrated Modular Avionics — A View on Safe Partitioning

Mike Ainsworth and Alan Simpson

Praxis Critical Systems Limited
20 Manvers Street, Bath BA1 1PX, UK
Tel: +44 1225 466991 Fax: +44 1225 469006
Email: ma,ajs@praxis-cs.co.uk

Abstract

Avionics systems are conventionally designed using a federated architecture. The system is partitioned into a number of physically separated black boxes, each of which is then treated as a separate subsystem. Integrity requirements are often defined at this level of resolution, with each black box being given an integrity requirement based on its most critical function.

In integrated modular avionics (IMA) architectures, there are no obvious black boxes. Although the basic principles of safety engineering can be applied to IMA, we believe that much of the detailed methodology of safety engineering for such systems will need to be modified. In particular, we advocate an approach which requires a much more detailed classification of safety requirements than is typically used in federated systems. Rather than regarding the system as a set of black boxes, we look at the internal workings of the system and use this detailed information in the development of the safety case — by analogy to software testing, we refer to this approach as "white box safety".

1 Introduction

Integrated Modular Avionics (IMA) is a proposed new architecture for future avionics systems. Instead of bespoke hardware with closely integrated software, systems will be developed to use standard hardware modules, connected in an integrated manner, with software interchangeable between modules of the same type. The major requirements that lie behind this architecture are outlined below.

One of the major problems faced by current military systems is that, due to the rapid frequency of change in commercial electronics, hardware (particularly microprocessors) is often obsolete before a system has entered service. This makes maintaining a system over its operational lifespan expensive and difficult.

IMA is intended to avoid these problems by making use of architecture independent programming techniques. Software will be designed to use an abstract interface to the hardware, allowing the actual hardware to be modified and upgraded without the software needing to be rewritten.

Another problem in avionics systems is that the use of bespoke hardware means that a large variety of spare components are needed, and providing fault tolerance in the aircraft often means duplicating entire systems. Failure of one component can ground an aircraft until the appropriate unit can be replaced.

An IMA system consists of a large number of interchangeable modules, all connected together. Since only a limited set of modules is available, management of spare units becomes much simpler. The connections between modules are reconfigurable, and since software can execute on any module, fault tolerance can be provided by a much smaller number of spare units. This has major benefits both in reducing the weight and space required for the avionics systems, and reducing the costs associated with maintenance.

1.1 Hardware Architecture

The hardware architecture of IMA can be divided into two main categories: core processing and interfaces. The following information is based on the description given in [1].

1.1.1 Core Processing

The core processing of the system is made up from a number of line replaceable modules (LRMs). These modules are physically mounted in a number of racks, and are connected by a switched optical communications network. The currently envisaged modules are:

- *Data Processing Module* — A general purpose processor unit, with some local memory.
- *Signal Processing Module* — A programmable signal processor for analogue and digital signals.
- *Power Conversion Module* — Supplies low voltage power for the various processing modules from the aircraft's main power supplies.
- *Network Switch Module* — Provides the network connectivity between the modules and racks.
- *Database Storage (Mass Memory) Module* — Storage for large volumes of data (cartographic information, imagery etc.)

1.1.2 Interfaces

A number of different interfaces are used throughout the aircraft. These are all connected to the core optical network.

1.2 Software Architecture

The currently proposed software architecture, based on information in [2], is as follows (see Figure 1):

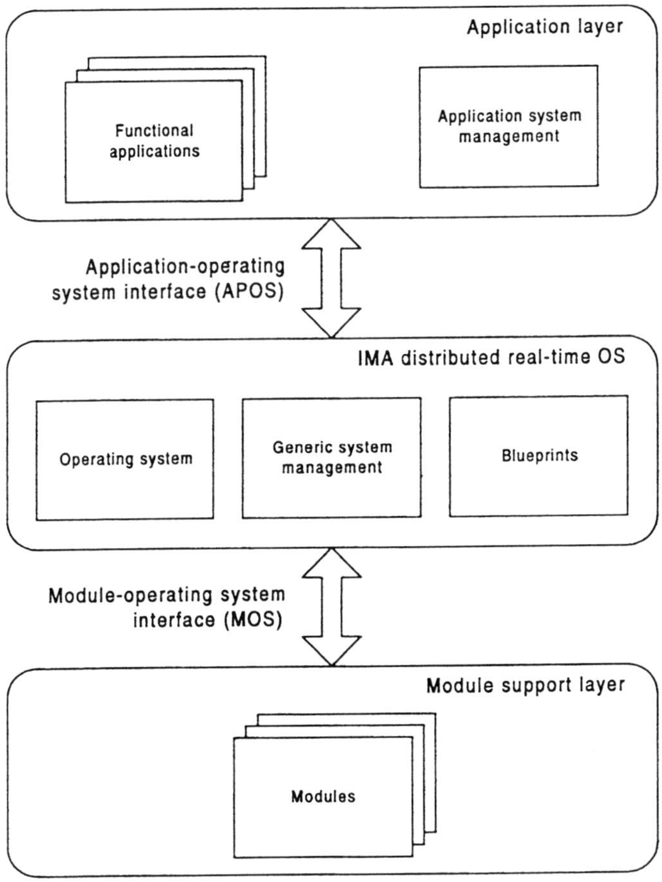

Figure 1 : IMA Software Architecture

- *Functional Applications* — The functionality of the aircraft or mission, e.g. flight control etc. The system will support multiple applications running on a single processing module. An application may also consist of several distinct processes.

- *Application System Management (ASM)* — ASM provides the interaction and control between the functional applications and the generic system management. This may involve reconfiguration requests, depending on phase of flight or mission scenario (which also include pilot interaction).

- *Generic System Management (GSM)* — This is the generic part of system management which uses information from the run-time blueprint and commands/interaction with the ASM to provide policy decisions at run-time.

- *Operating System (OS)* — The OS is a defined set of services which provides:
    - the implementation of policy decisions from the generic system management;
    - the services to the layer above, e.g. communications through the APOS interface.
- *Run-Time Blueprint* — The Run-Time Blueprint is a repository of information which the GSM uses to tailor its operation. This must be co-ordinated with the ASM component implementation.
- *Module Support Layer (MSL)* — The MSL consists of a set of low level, hardware dependent, services providing basic functionality to the layer above through the MOS interface.

1.3 Blueprints

A system as flexible as IMA can clearly be configured in many different ways. Calculating the best possible configuration based on the currently available modules and the mission requirements at the time is a difficult process. Consequently, IMA uses the concept of *blueprints*. These are possible system configurations which are pre-calculated and stored within the system. This means that at run-time the system only has to choose between a limited number of pre-defined alternatives, rather than performing a complex optimisation calculation.

2 Partitioning

Partitioning means dividing up the resources of a system so that each process has the resources it needs to execute correctly. This minimises the potential for undesired interference between processes and thus reduces the complexity of the system.

The requirement to demonstrate specific integrity levels in a safety-related systen forces partitioning of the system in order to ensure that the probability of a failure in one component resulting in an accident is reduced to a tolerable level.

The precise nature of the partitioning will vary according to the type of resource being partitioned and the types of interference we are most concerned with avoiding. It can range from purely theoretical partitions to physical separation. These issues are discussed in a general way in the following sections.

2.1 Partitioning Problems

Although the precise nature of potential interactions between two processes will depend on the nature of the shared resource, it is possible to provide some general classifications of errors, and use these to guide the design decisions.

2.1.1 Direct Interference

2.1.1.1 Spurious reads

One potential class of error is where one process can access a resource assigned to another process and read information from that resource. In general, reading data from a safety-related process is unlikely to be a safety problem per se, although it may lead to other problems, such as starvation (see below). However, unauthorised reading of data would be a problem for processes which contained protectively marked information.

2.1.1.2 Spurious writes

A more serious class of access problem is where an unauthorised process can overwrite data in a critical process. If the potential for this sort of error exists, then predicting its effects on the behaviour of the critical process becomes extremely difficult, with consequent problems for certification.

2.1.2 Indirect Interference

The above problems are both instances of direct interference, where one process directly reads or writes into another processes resource area. On the whole such problems are fairly easy to identify at the design stage, and partitioning can be applied to prevent them in a relatively straightforward way.

More subtle problems can arise as the result of indirect interference. A few examples are described below, but it should be noted that even when components are operating entirely within their specification, it is still possible for them to create indirect interference.

2.1.2.1 Starvation

The simplest from of indirect interference is starvation, where a process may only be accessing its own resource area, but does so at a sufficiently high rate that other processes are prevented from accessing their resources.

2.1.2.2 Deadlock and Livelock

It is possible in a concurrent system to get into a situation where two processes both obtain part of the resources they require, but are unable to meet their complete requirement due to the resources being used by the other process. At this point the processes are deadlocked — neither process can proceed.

Livelock is essentially an infinite loop involving a group of processes. Although each process is executing, no useful progress is made. This situation often arises in communication protocols, e.g. if a process fails to receive an acknowledgement, it may continually retransmit a message.

2.2 Partitioning Techniques

Although the precise nature of the partitioning will depend on the type of resource being partitioned, it is possible to use the following general classification scheme to discuss the properties of different partitioning techniques.

2.2.1 Separation in Time and Space

Resources can be separated in space or time:

- Separation in space means using different resources;
- Separation in time means using the same resource, but at different times.

In general, separation in space gives a higher level of protection, but at the increased cost of duplicating resources. Separation in time gives very efficient use of resources, but ensuring availability and fairness may be more difficult.

2.2.2 Physical and Virtual Segregation

Partitioning varies from complete physical separation through to virtual (i.e. software) techniques. There are four main levels of partitioning that can be used:

- Physical segregation means that it is not physically possible for two processes to access the same resource;
- Active segregation means that while two processes may share a physical resource, a third process actively polices the system to enforce separation;
- Passive segregation means that the processes share a resource without explicit separation, but implement a protocol to keep their usage separate;
- No segregation means that processes are competing for the resource without any constraints.

This may become clearer if we consider how these different types of partitioning might be applied to a network (see Figure 2):

	Physical Segregation	*Active Segregation*	*Passive Segregation*	*No Segregation*
Space	Separate networks	Firewall	Bridged LANs	Single network
Time	Circuit switching	OS Scheduling	Timing models	Unpredictable

Figure 2 : Partitioning Levels for a Network

In general, separation in space is stronger than separation in time, although less efficient in resource usage.

2.2.3 Properties of Different Partitioning Techniques

2.2.3.1 Passive Segregation

A protocol-based partitioning scheme will usually be able to detect faults, and in some cases may be able to prevent direct interference, but offers little protection against indirect effects.

If a sufficiently complex protocol is used, then the probability of another application accidentally generating a valid message is sufficiently low that it can regarded as extremely unlikely to occur. However, this can add a lot of overhead processing, and would be difficult to implement for resources which require very high speed transactions.

The problem with a passive scheme is that although it can allow interference to be detected, this only happens *after* the message has been transmitted. This means that a certain amount of time and resources will be used in handling invalid messages. If the resource includes some level of buffering (e.g. a network) then the system will be able to cope with some level of spurious faults, although a high rate of faults might be sufficient to cause a genuine request to be missed. If the resource is unbuffered (e.g. memory), then a single invalid transaction might cause unrecoverable corruption of data.

2.2.3.2 Active Segregation

An active segregation scheme is considerably stronger than a passive scheme; in particular it can prevent direct interference and provides some level of protection against indirect interference. For example, a network firewall would be able to prevent spurious messages from being transmitted across the network. Although spurious data might be generated, it would be confined within a particular firewall partition, and thus would be limited in which systems it could affect.

However, there are two outstanding problems with such schemes:

1 It is still possible to communicate through the partition, albeit in a strictly limited fashion;

2 We need to validate the partitioning agent.

The first of these means that although it is unlikely that all forms of interference across the partition would be eliminated, it should be feasible to reduce the possible forms of interference to a level where they can be handled by other analysis techniques.

The second problem is more fundamental. Although a number of commercial partitioning devices exist for various types of resource, none have the level of validation required for use in safety-critical systems. It may be possible to apply a similar argument to that used for microprocessors, and argue that a commercial device will have had sufficient hours of use that any deficiencies will be already known. Alternatively, it may be feasible to design and validate devices specifically for IMA applications, although this would be a major task.

2.2.3.3 Physical Segregation

Physical segregation is the only partitioning scheme that is strong enough to be used to demonstrate non-interference (including absence of indirect interference) in a straightforward manner. However, physically segregating an IMA system would limit the potential for reconfiguration (although there would still be some logistic savings due to the use of common modules).

The decision as to whether physical segregation is required has to be driven by the type of safety requirements that must be met, and the required integrity level for each requirement. For some types of safety requirement, a physical segregation scheme may be the only way to demonstrate the requirement to the required degree of rigour.

3 Certification

The term *certification* in military systems usually refers just to assessment of the safety of a system. There is a similar but separate process applied to the assessment of the security of a system. Although this paper concentrates on safety integrity, we suggest that, in a system as integrated as IMA, it will be necessary to consider integrity in a more general sense, encompassing safety, security and reliability.

3.1 Safety

The basic requirement for safety is to assess the hazards that might arise during the lifetime of the system, and then to show that the risk of each hazard occurring has been reduced to as low as is reasonably practicable.

To achieve this it is necessary to be able to define the states of the system with higher risks, and give arguments about how the system can move between states. The aim here is to demonstrate that the probability of entering a high-risk state has been minimised, and that the time spent in any high-risk states has been minimised.

When a fault is detected in a safety-related system, the fault needs to be handled in a way that minimises the risks. This fault handling will necessarily be application specific — in general the system needs to move to a state with a lower risk factor, but in some continuous demand systems there might not be a single 'safe' state that the system can fail into.

3.2 Demonstrating Requirements

Certification requires a demonstration that the system meets its integrity requirements, this is based on a combination of the application of rigorous development techniques and the evidence obtained from vertification and validation (V&V). These activities are supported by independent audit and assessment.

It is important to be clear about what precisely is being certified. Certification always applies to a complete system (e.g. an aircraft) with a specified fit of software and hardware, performing a particular operational profile. Software cannot be certified without reference to the specific hardware on which it executes.

Certification therefore requires a detailed definition of the complete environment in which a particular software item executes. The configuration management of such systems is becoming increasingly more complex as systems get larger and more COTS components are used. For example, with a microprocessor, faults may be present in one particular revision of the chip but not in others.

3.3 Certification and IMA

IMA includes a number of novel aspects which will raise new problems for certification.

In a conventional federated system it is usually trivial to demonstrate that most of the equipment on the aircraft cannot possibly influence the environment of the safety-related software under consideration. This means that the environment in which the software operates is essentially fixed, and so the certification process only needs to consider a relatively constrained portion of the aircraft.

In an IMA system it is much more difficult to isolate the environment in which a piece of software will operate. The connectivity and the potential for interaction in IMA means that the environment is much more fluid and so there are potentially far more scenarios which must be considered for certification. In addition, a particular processor module may contain software which has different Safety Integrity Levels and which may have different security requirements.

IMA systems are also likely to be upgraded on a more frequent basis than conventional systems. This means that the certification of the aircraft will need to be repeated for each modification.

In order for certification of IMA systems to be practical, we must avoid a situation where all the software must meet the integrity requirements of the most critical application, or where everything must be re-certified every time a modification is made.

In order to achieve these goals, certification of an IMA system will need to be based on:

- Certification of a specific aircraft subsystem, rather than the whole aircraft. The subsystem will be precisely defined in the blueprint, i.e. software and required resources. The blueprint will need to include details of any hardware or software which might result in a hazard if modified.

- A safety case which demonstrates that the subsystem is safe provided that the conditions laid down in the blueprint are met.
- A system architecture which provides sufficiently rigorous partitioning that we can exclude large portions of the software as potential causes of hazardous failure. So if the blueprint says that an application only communicates with two other processes, then those are the only processes which it is allowed to communicate with.
- An operating system which can be depended upon to meet its specification to a very high level of integrity.

This approach allows certification authorities to gradually move from the current certification approach (certify the aircraft) to a more component-based approach as a gradual process. As a first step, if the blueprint includes a precise specification of *all* the hardware and software on the aircraft then the certification would be identical to the present situation — if anything is modified, the subsystem must be re-certified.

4 Designing IMA Systems

There are a number of issues which will need to be considered when designing IMA systems.

4.1 Safety Engineering

4.1.1 Hazard Analysis

The way in which 00-56 is currently interpreted on many projects tends to result in "SIL4 black boxes". This concept of *black box safety*[1] tends to lead to two major problems when developing and certifying systems:

1. Initial hazard analysis sometimes lacks detail, resulting in difficult rework later in the project when previously unnoticed hazards are discovered;
2. There is a tendency to demonstrate far more properties to SIL4 than are required just to prove the safety requirements. This is one of the major contributors to the 'large pile of paper' approach to the safety case.

[1] We use the terms *black box* and *white box* for safety engineering in an analogous way to black box and white box testing — where black box testing means testing without making assumptions about how the box works, and white box testing means testing based on detailed knowledge of the internal structure.

These problems already occur in some existing federated developments — IMA is likely to exacerbate the situation.

What we would like to suggest is moving away from this black box approach, and using an approach of *white box safety*, where we look at the details of the functionality, and concentrate on proving specific safety properties, rather than applying blanket verification activities. The emphasis is to focus our attention on error modes which may result in a hazardous failure, and to rationalise about errors which only have a marginal impact on safety.

Safety requirements define what must be proved. *Integrity levels* define how rigorous that proof has to be. Integrity levels should be assigned to specific requirements or functions, not to systems.

This does not change the design process, but it will have a major impact on verification and validation. Current projects tend to apply blanket levels of verification and validation to all components in a system. Although some level of functional correctness will need to be demonstrated as part of any development, the very high levels of rigour required for safety-related functions should be applied in a very targetted way.

4.1.2 Safety Analysis in a Concurrent System

Hazards are associated with functional failure or physical equipment, so changing the software architecture without modifying the functionality cannot introduce any new hazards into the system. What IMA may affect is the causes of hazards and the probability of a particular hazard developing.

The fact that IMA uses a concurrent, distributed operating system means that faults (deadlock, livelock etc.) can occur in an IMA system as a result of concurrency which cannot arise in a conventional system.

Conventional safety analysis techniques are usually designed to describe situations arising from a single fault or a combination of faults. They often have problems describing the complex situations involved in concurrency problems, where there is no single fault, but rather a complex pattern of behaviour involving several different processes.

Praxis Critical Systems was involved in developing guidance material for the nuclear industry on the use of communicating concurrent processes in safety-related systems [3]. Much of this guidance would be applicable to IMA, however the two key points for safety analysis are:

- It is possible to adapt techniques such as HAZOPS to identify hazardous concurrent failures, however, this requires an understanding of concurrency issues on the part of the safety engineer.
- It is almost impossible to identify concurrency problems using testing. The use of formal verification techniques such as model checking is strongly recommended.

Where concurrency is used as part of the software design it is important that the specification techniques and programming languages used provide appropriate features for describing concurrent behaviour (e.g. Ada tasks).

4.1.3 Types of Safety Requirement

It is common practice whilst developing safety-critical systems to concentrate on the integrity level to be assigned to subsystem components, rather than derive the safety properties of each function. Although a group of functions may all have the same integrity level, they can have fundamentally different qualities to their safety requirements. In particular, it is useful to distinguish between protection functions and continuous control functions.[2] The differing nature of these requirements has a considerable impact on the ease of integrating them into an IMA environment.

4.1.3.1 Protection Functions

Protection functions monitor other system controlling functions. This generally means that the safety requirement is to prevent the system performing some hazardous action. These are typically expressed in terms of describing what the system must *not* do. Such systems always have a safer state into which they can fail, typically by *not* doing anything, or by triggering an alarm. The safer state needs to be enforced by the actuator (the hardware being controlled) rather than the IMA application, since it needs to remain safe even in the event of a systematic failure of the IMA operating system.

In general if the capability to perform a protection function is lost, the aircraft will still be safe, although the system will lose whatever capability the safety function was protecting, which may be unacceptable in mission terms.

In terms of partitioning, with a protection function the partitioning needs to be demonstrated around the controlling application and the actuator (and around inputs where necessary), but it is not generally necessary to prove results about the communication between these partitioned zones.

4.1.3.2 Continuous Control Functions

A continuous control function is one that has to be available in order for the system to remain safe, although the function may only be required during certain operational modes (e.g. in flight). Typically this may be to ensure that the system remains in a safe state in response to dynamic changes in the environment, or to provide some functionality which is needed in response to an emergency. Requirements for control functions are usually described in terms of what the system *must* do. Control functions often need to respond to data in real-time in order to

---

[2] In computer science, these functions are referred to as *safety* and *liveness* properties, where a safety property describes what the system must *not* do, and a liveness property describes what the system *must* do. We use the terms protection and control functions since having two different meanings for 'safety' is rather confusing.

maintain safety, and although there may be scope for some graceful degradation of service, there is usually no simple fail-safe behaviour.

If the capability to perform a continuous control function is lost, the aircraft will be unsafe. There is usually no way to continue operation of the aircraft if such a function is lost, so such functions require very high reliability.

For continuous control functions, the partitioning needs to form a contiguous channel between the inputs, controlling application, and actuators.

4.1.4 Safety and Reliability

Where a system has a safe state, it is entirely possible to build a safe system from less reliable components, as long as the designers ensure that any fault will cause the system to fail in a safe manner. For these systems, safety and reliability are fundamentally different quantities, and it is perfectly possible to have one without the other. In fact, making the system safer often tends to decrease its reliability, because it is better to fail in a safe way than add complexity in order to avoid the fault.

For control systems, where continued functionality is required to ensure the safety of the aircraft, safety and reliability are inextricably linked. Although it may be possible to design the system in such a way that minor faults can be tolerated, an overall system failure is intolerable.

4.1.5 Example

Consider the example of a stores management system, which might possess the following hazards:

- *spurious launch* — a store is launched without proper authorisation (i.e. no command from the crew, or the launch is not safe at this time).
- *failure to jettison* — the crew have commanded a store to be jettisoned, but the store is not released.

The launch function is an example of a protection function, it has a safe state, and the safety requirement is concerned with preventing particular actions. The jettison function is an example of continuous control — it must be available at all times and must occur when requested.

It is important to distinguish the level of tolerability associated with each hazard. In this case, we might observe that a failure to jettison would only result in an accident if the aircraft was already in a hazardous state (which itself is a low probability event). A spurious launch, on the other hand, might result in an accident if it occurred when the aircraft was carrying live stores (a reasonably high probability scenario). This would imply that the tolerability of a jettison failure is much higher than the tolerability of a spurious launch, with a corresponding impact on the designated integrity levels.

4.1.6 Systematic and Common Cause Failure

In general, for most safety-related systems it is impractical to reduce the probability of any single fault occurring to a level which meets the required safety integrity. Instead a safety engineer tries to reduce the probability of a hazard developing by ensuring that a hazard can only arise as a result of multiple independent failures, rather than a single fault.

Although duplicating an application across multiple processing sites would reduce the risk of a random hardware failure causing a hazard, we could still face problems due to a systematic failure or a common cause failure:

- *systematic failure* — a design fault in the software or hardware might cause each instance of the application to fail in the same way.
- *common cause failure* — a fault in the underlying operating system or network could affect all instances of the application.

Systematic failure in IMA can be addressed in exactly the same way as for a conventional system (i.e. rigorous design methods and verification techniques), however there may be new problems to consider (e.g. concurrency).

Common cause failure due to a problem in the underlying system architecture will be a major concern for IMA. This is likely to mean some degree of physical partitioning to provide mitigation against an operating system failure.

4.1.7 Risk Reduction

Having identified the potential causes of a hazard, the safety case would then have to demonstrate that the risk of the hazard occurring had been reduced to the lowest practicable level. Partitioning could be used either to reduce the risk of a fault leading to a hazard, or to reduce the complexity of the system so that it is easier to demonstrate the required safety properties.

For example, one concern for a launch function might be another application sending a spurious launch message. The probability of this occurring could be drastically reduced by adopting network protocols which included verification of the message's source, destination and contents.

In order to reduce the risk of a fault within the stores management application causing a spurious release, the application (or the relevant part of it) might be duplicated and both copies required to simultaneously request a launch before the store is released. This same strategy could be used to reduce the risk of the stores management application failing to request a jettison, by having two duplicates and allowing a jettison if either of them sends a message to the stores interface.

4.1.8 Separation of Safety-Related Functionality

In order to keep the problems of certification to a reasonable level it is necessary to clearly separate any safety-related functionality.

The partitioning techniques within the IMA operating system provide a technical mechanism to do this, but the use of these partitioning techniques must be considered as part of the application design.

In particular, within a safety-related application there is typically only a very small proportion of the functionality that actually has specific safety requirements. If the small portion of the application which is actually safety-critical can be partitioned into a separate process, the certification of the system will be much easier.

4.1.9 Targetted V&V

If the safety-related functionality can be clearly identified then it is possible to be far more selective in our approach to verification and validation. Rather than trying to prove the complete functionality of an application to a uniform level we suggest an approach which looks at the specific types of fault that might arise and whether those faults might result in a hazard, and then tailors the V&V to demonstrate specific properties.

- If a fault can result in a hazard, then it is necessary to demonstrate the absence of such faults to the required integrity level for the software.
- If a fault cannot result in a hazard, but may result in unreliable performance, then it is necessary to demonstrate the absence of the fault to a level which meets the reliability requirements of the software.

By distinguishing between hazardous and non-hazardous failures the most rigorous and time-consuming aspects of V&V can be restricted to just those areas where they are needed, rather than applying a uniform level of V&V activity to the complete system.

This approach means that the application as a whole will have been demonstrated to meet its safety requirements to the required integrity level. However, non safety-related requirements may not be demonstrated to the same level of rigour.

4.1.10 Example

The software for the stores management system will probably include some built-in-test functions. A black box approach to the verification and validation would take the view that such functions formed part of a SIL4 unit, and hence should be validated to SIL4 by applying rigorous analysis techniques.

A white box approach would ask how these parts of the software might lead to a hazard. In most cases the functionality of the BIT routine is irrelevant to the hazards, and the real safety concern is whether a fault in the software could somehow overwrite a variable elsewhere in the system which was concerned with weapons release. Rather than trying to prove the full functionality of the BIT routine to SIL4, the most rigorous techniques might only be used to prove the absence of memory leakage problems, and a lower level of rigour applied to demonstrating the functionality.

4.2 IMA Architecture and Operating System

4.2.1 Dependability

It is possible to implement protection functions using unreliable components, since the system can be designed to fail into a safe state (although such a system would still have to ensure some minimal level of reliability to meet the mission requirements). However, to implement continuous control functions a dependable architecture is required.

Re-validating the operating system for each new application would clearly be impractical. This means that it must be possible to argue rigorously about the behaviour of the operating system without detailed investigation — it must be possible to trust the operating system to meet its stated functionality.

To demonstrate that a continuous control function meets its safety requirements the designers must be able to show liveness properties, and this implies that the IMA architecture must include the following in its requirements:

- the operating system must be free of deadlock, livelock and starvation;
- all operations must complete within a stated finite time;
- all operations must have a formal specification of their functionality, including timing properties.

To stand any chance of being able to prove these properties to the required integrity levels the system must have a formal specification, and the programming language used must support rigorous proof (e.g. SPARK Ada).

4.2.2 Partitioning

In order to exclude faults in other applications as potential causes of failure, it is necessary to demonstrate that other applications cannot interfere either directly or indirectly with a safety-related application in such a way as to cause a hazard.

This requires either:

- active segregation of all shared resources using a partitioning agent which has been validated to the same level of integrity as the safety-related application, or
- physical segregation.

Passive segregation is generally insufficient for such purposes.

For some safety-related applications it may be feasible to show that certain classes of fault cannot result in a hazard, and so it may be permissible to reduce the partitioning requirements. However, since this can only be done on an application specific basis, the IMA system must provide facilities for rigorous partitioning of all resources. In some cases the nature of the hazards is likely to mean that physical segregation will be required.

4.2.3 Common Cause Failure

Failures within the operating system could affect any application using the IMA system, and create the potential for a failure to affect multiple critical systems. Such failures will therefore be a major concern in the development of safety-related IMA systems.

There is a finite limit to the level of integrity which can be demonstrated for the IMA operating system. This means that even with the best possible development practices, there will still be a small probability of an operating system fault, and the probability of such a fault being encountered will be proportional to the number of applications using the operating system — more applications running means a higher risk of an operating system fault being encountered.

To reduce the risk of such an operating system fault affecting a safety-related application there must be a finite limit to the number of applications using the operating system at the same time. This limit means that the overall IMA system must be partitioned into a number of distinct pieces, and this partitioning must be done at a level below that of the operating system.

This implies that the overall IMA system must be physically partitioned into a number of distinct pieces to prevent an operating system fault leading to an unacceptable common cause failure. The precise rules for when such partitioning must occur will depend on the number and integrity level of the applications using the system.

Since physical partitioning would reduce the potential benefits of IMA, one possible solution would be to only apply physical separation to very specific safety functions.

4.2.4 Safety Management

Whenever a new application is added to an IMA aircraft, the designers must either demonstrate that it will have no impact on safety, or they must provide a safety case for the application.

Providing a mechanism for recording safety requirements as part of the blueprint will considerable ease the process of certifying upgrades to the system. For certification of such a complex system it is vitally important to be able to distinguish between functional requirements and safety requirements.

5 Conclusions

Conventional systems are designed as a series of physically distinct "black boxes", and the safety certification relies on this partitioning as a way of limiting consideration to a small part of the aircraft.

In IMA there are no black boxes, and even where there is strong software-based partitioning between applications, the situation is likely to be far more complex than a conventional system.

Current practice in many projects is to designate an entire physical unit as "SIL4" and then apply uniform development and verification techniques to everything within that unit. If such an approach is followed for IMA, certification is likely to become unmanageable.

Instead we believe a more detailed approach is needed in which integrity levels are assigned to specific functions, and safety requirements are separated from general functional requirements. We still prove the software to be "correct", but the level of rigour is varied according to the safety impact of a particular requirement. Rather than trying to prove the complete absence of errors, specific faults which might result in a hazard are identified, and the absence of those faults demonstrated. Such an approach allows a more tailored technique for verification and validation to be adopted, in which the presence or absence of specific properties is demonstrated, rather than full functional correctness.

This approach is already being used in a limited way on conventional systems which employ software with differing integrity levels executing on a single processor. It is possible to justify applying differing levels of verification to the software on such systems, but it requires a detailed understanding and safety analysis of the software and its possible failure modes.

We refer to this approach as "white box safety" because it exploits detailed knowledge of the internal workings of the system in the production of the safety case.

6 Acknowledgements

This paper is based on the results of a research study undertaken by Praxis Critical Systems Limited on behalf of the Defence Evaluation and Research Agency under contract ARP 07c/61. The paper does not represent the views or policies of the Ministry of Defence. We would like to thank DERA for giving their permission for this work to be published, and to the various reviewers within DERA and Praxis who have contributed to the project.

7 References

1. R A Edwards, *Exploiting the Potential of Integrated Modular Avionics,* Proc. 1997 Avionics Conference and Exhibition, November 1997.

2. D Field et al., *Software Techniques for IMA System Management, System Security and Blueprints,* Proc. 1997 Avionics Conference and Exhibition, November 1997.

3. *Guidance on CCP Safety,* Report number S.P0545.40.1, prepared for HM Nuclear Installations Inspectorate by Praxis Critical Systems, October 1997.

Independent Verification Magic or Myth?

Andrew J Nolan
Software Engineering & Assessment Services: Rolls-Royce
Derby, England

Abstract

The author challenges the roles and responsibilities of the developer and independent verifier during testing. Traditionally the independent verifier would design and perform the tests. The author proposes that, instead, the developer should design and perform the tests and that the results are reviewed independently. The author questions what independence is providing and explores whether a closer collaboration between independent verifier and developer could produce a more cost-effective testing process.

1 Objective

However difficult, the dream is to achieve zero errors in a delivered system through verification and validation. The paper proposes a way to more effectively achieving this dream and to do this at reduced cost. The scope of the paper is aimed at real-time safety-critical software (SIL 3 or SIL 4 as defined by IEC 61508 [IEC-61508 1998]), although the principles may apply equally to other types of software and integrity levels.

This paper addresses the concept of independence during verification as defined by many standards and current practices. At present, the "developer" would design and build a system. The "independent verifier" would review and test the system. "Independent" may mean a different person, department or organisation depending on the integrity level of the software and the standards followed.

There are many reasons why independence is required. An independent verifier can easily spot the mistakes made by the developer. The independent verifier can spot misinterpretation of requirement, which the developer would otherwise carry through into testing. The independent verifier is not always subject to the same project pressures and therefore can perform a more objective test. In addition, the independent verifier can break a system that the developer may find psychologically difficult to break. The paper is not in dispute with these benefits and it proposes to continue to use a degree of independence during verification. However, what the paper does question is the appropriate balance of the roles and responsibilities of both the developer and independent verifier during testing.

For many organisations, testing is performed independently. It is proposed that, instead, the tests be designed and carried out by the developer and that the results and testing strategy are reviewed independently. If the independent verifiers find deficiencies in the tests performed then these will be flagged to the developers and further tests agreed and performed.

Testing can account for up to 50% of the overall project costs. Any improvement in the effectiveness of this activity would have an enormous impact on the overall cost of the project. The paper is written to reflect industrial experiences gained by the Software Engineering and Assessment Services (SEAS) department of Rolls-Royce, through many years of developing systems and providing an independent verification service. Some of SEAS's experience, presented within this paper, goes against the common beliefs about independent verification and show that, in some cases, the developer can perform a more significant role during verification without compromising independence. Through arguments and examples presented within this paper, it is suggested that the change of roles and responsibilities will increase the quality of the testing process and at the same time, reduce the overall project cost.

The paper refers to "testing" as the dynamic testing activities of unit testing, integration testing, system testing and target testing. Although the focus is predominantly on testing, the same principle can apply to some aspects of static analysis including data flow, control flow, information flow and semantic analysis. Acceptance testing would still be performed by the customer and static reviews and safety analysis would still be performed independently.

2 Understanding the Process

A "friend of a friend" once described how his wife would slice the ends off the joint of beef before roasting it. His wife's explanation was that slicing the ends off improved the flavour of the beef. The source of this pearl of wisdom was her mother. Eager to get to the source of this behaviour, he approached her mother and again asked why she sliced the ends off the beef. The same answer was given - it improved the flavour of the beef! The source of this advice came from the grandmother. On asking the grandmother, she replied "I cut the ends off the beef because my baking tray was only *this big*, and I had to cut the ends off to get it to fit in".

If anything, this story illustrates how behaviour or good practice can lose its meaning over time. Like Chinese whispers, the message is lost through reinterpretation, misinterpretation or simply a lack of understanding. What was once a good idea and appropriate for the problem then, may no longer be appropriate for the problem now. When reasoning is lost, we often rejustify our actions with unscientific explanations. As in this example, it was believed that slicing the ends off the beef would make it taste better. The question that needs to be asked is just how much of our working lives do we carry out tasks without question, which are no longer appropriate for today.

An "excellence modelling" study was performed into software best practices. The intention was to study the best developers and to capture their processes, strategies, beliefs and behaviour. Although there have been many such studies to capture best practice, the study was repeated with a control group of less capable developers. The intention was to identify not only what the best people do, but also what they do that others do not - the difference that makes the difference. The single most distinctive difference between the success group and the control group was that the success group could identify what was most appropriate because they understood WHY they did things.

It is the successful group's understanding about what was most appropriate that allowed them to be more flexible. This principle led to life cycles that were more productive because the developers understood which processes were most appropriate for the project and therefore reduced or removed nonessential processes. It produced higher quality code and designs because developers understood why it was important. It also produced better customer relations because the developers changed their communication approach appropriately for each customer.

If we are to remain efficient and effective in our work, we must understand why we are doing activities so we can choose those actions and activities that are most appropriate for what we are trying to achieve. Many software methods and techniques of today were developed in the sixties and early seventies. However, systems, tools and languages have improved dramatically since then. The methods that were appropriate in the sixties may no longer be appropriate for today.

3 What the Standards say About Independence

The definition of roles and responsibilities, and the need for independence, is driven from the standards. Table 1 illustrates the degree of independence required by each standard for the range of integrity levels, from low integrity to high integrity reading down the table. Against each standard and integrity level are ticks showing the level of independence required between the developer and verifier. In some instances the developer may perform their own testing whilst for higher integrity levels independence may be a separate person, department or organisation.

From table 1, it is possible to observe a range of requirements for independence between standards for the same level of software integrity. This raises the question that if independence is so critical, why does it vary between the standards. What is the most appropriate level of independence? As the standards differ, does this suggest that we have not found an appropriate level, or that it has been found, and it differs appropriately for each application domain?

The proposal made by this paper is that we should allow the developer to perform their own tests, and achieve independence during the review of the test specifications, test results and static analysis reports. The quotes below are taken from DEF STAN 00-55 [00-55 1997] and indicate that this is a plausible solution, although only for low software integrity levels:

> "Where the testing is designed and conducted by the Design Team, the Test Specification shall be reviewed by the V&V Team" DEF STAN 00-55 Part 1: clause 37.1.4

> "When the testing is designed and conducted by the Design team the testing process shall be audited by the V&V Team." DEF STAN 00-55 Part 1: clause 37.1.5

Standard	Integrity: low to high	developer self testing	Independent Person	Independent Department	Independent Organisation
IEC 61508 Generic Standard [IEC-61508 1998]	SIL 1		✔		
	SIL 2		✔	✔	
	SIL 3			✔	✔
	SIL 4				✔
EN 50218 Trainborne Systems [EN-50128 1996]	level 0	✔			
	level 1		✔		
	level 2		✔		
	level 3		✔		
	level 4		✔		
DEF STAN 00-55 Defence Systems [00-55 1997]	S1	justify	justify	✔	
	S2	justify	justify	✔	
	S3			✔	
	S4			✔	
DO-178B Airborne systems [DO-178B 1992]	D	✔			
	C	✔			
	B	✔	✔		
	A		✔		

Table 1 : Level of independence between developer and verifier

Most developers, coders and designers will perform a range of self-testing activities even if this is only to compile code or spell check and grammar check a report.

Many developers carry out more invasive tests on their code by running several test cases through the compiled program. Although informal self-testing is not a new concept, carrying out formal self-testing activities for high integrity software is a change in current practice and not generally supported by the standards.

4 Why do we have Independence?

The four quotes by Myers [Myers 1979] below illustrate the importance of independence:

> "...testing can be viewed as being analogous to proof-reading or writing a critique of a paper or book. As many writers are aware, it is extremely difficult to proof-read and critique one's own work."
>
> "...the organisation responsible for developing a program has difficulty in objectively testing the same program....it is desirable that the test organisation is not be part of the same company, for if it is, it is still influenced by the same management pressures influencing the development organisation."
>
> "...it is extremely difficult, after a programmer has been constructive while designing and coding a program, to suddenly, overnight, change his or her perspective and attempt to form a completely destructive frame of mind toward the program. As many home owners know, removing wallpaper (a destructive process) is not easy, but it is almost unbearably distressing if you, rather then someone else, originally installed it...."
>
> "...the program may contain errors due to the programmer's misunderstanding of the problem statement or specification. If this is the case, it is likely that the programmer will have the same misunderstanding when attempting to test his or her own program."

Although Myers refers to programmers, the same arguments apply equally to any development activity. Carrying out all testing activities independently may resolve these issues but is it introducing other problems? For example, how many error types are there? How many of these errors can be found by the developer and how many by the independent verifier? Who is most effective at finding which type of error? If we do not have an answer to these questions, then we do not fully understand the costs and benefits that independence brings. How do we know that independence is appropriate, and how do we know when it is appropriate?

Experience and evidence gather by SEAS suggest that the developers are effective at finding some types of errors and, as described by Myers, misinterpretation of requirements can be found better by the independent verifiers. However, by using independent verifiers for all testing activities, are we attempting to find some error types at the potential risk of missing other types of error? If the developer is better at finding some types of error and the independent verifier at finding others, then clearly it is appropriate to use the best from both teams rather then using only one team or the other.

5 Case Studies

The following is a range of case studies illustrating the effectiveness of the developers performing their own tests. The case studies are all from SEAS. SEAS is Rolls-Royce's centre for high integrity and safety related software. Rolls-Royce formed SEAS to provide a centre of capabilities for software development, software verification and validation, consultancy and research.

An "Excellence Modelling" study was performed at SEAS on software verification. The intention was to study applications that we believed were developed efficiently and to a high quality. Those applications with no operational errors were considered as our success group. The study revealed many characteristics of the most successful applications. One factor was the use of self testing, either formally or informally. The case studies below are a sample of the projects studied. The case studies were taken from a range of projects and developed to a range of software standards.

5.1 7500 lines of Ada-83 written to SIL 2. The program was tested independently using AdaTest. 1700 hours were booked by the independent verifiers during testing. In parallel, the developers booked 600 hours testing the same system. The independent verifiers found two anomalies, the developers found 34 anomalies, including the two anomalies identified by the independent verifiers.

5.2 A scientific application of 48,000 lines of C developed to SIL 2. The application was statically analysed, unit tested and integration tested by the developer and acceptance tested by the customer. Developer testing and static analysis was performed as an integral part of the development process rather then as a post-design activity. Static analysis addressed data flow, information flow and control flow. All tests were independently reviewed by the customer for completeness and accuracy. During acceptance testing, only two errors were found which covered a number of minor spelling mistakes and a change to requirements.

5.3 6000 lines of Ada written to SIL 4. The application was informally tested bottom up during development and errors removed. No records were kept of the informal testing process. Over 3000 hours were spent independently testing the application and no anomalies were found. However, during independent visual code inspection two minor anomalies were found. The anomalies were suggestions to use 32-bit arithmetic rather then 16-bit, to ensure a greater accuracy during calculation.

In each of these case studies, the developers performed their own tests and the number of errors identified during independent testing was low. In addition, the developer testing was considerably faster then the independent testing. This suggests that in some cases, there is little benefit in all testing activities being performed independently. In one instance, independent testing was less effective than the developers testing their own system.

Although these examples conveniently illustrate the point made by the paper, there may be other factors at work that are affecting these results, and this pattern may not necessarily be common to all organisations. However, if there is even a single situation where the current roles and responsibilities during independent verification have been shown to be ineffective, we must reconsider what independence is providing and where it is appropriate to use it.

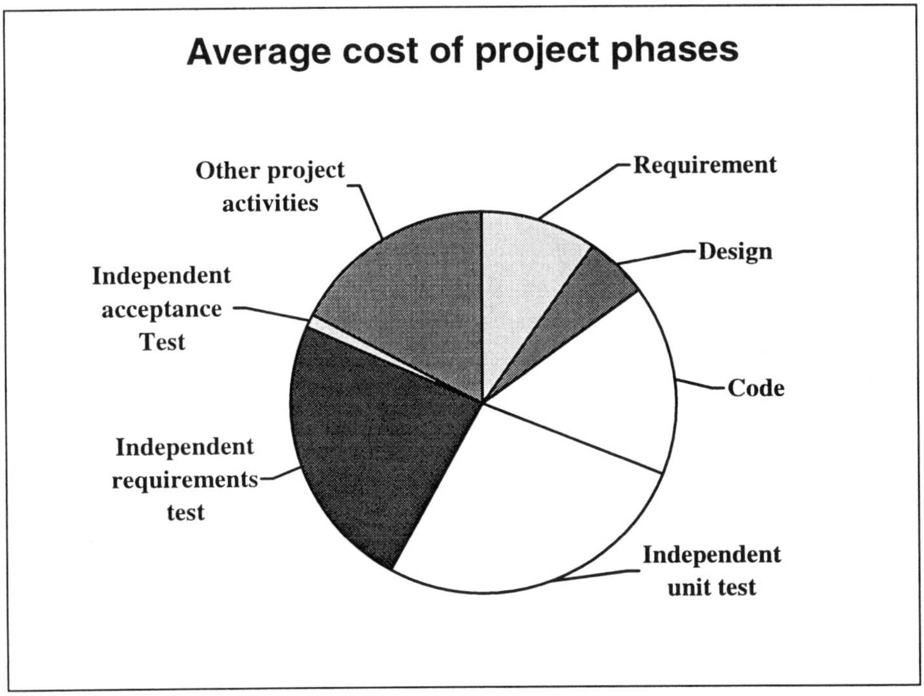

Figure 1: Average cost of project phases including independent dynamic testing

Myers suggests that it is psychologically difficult for a developer to test their own code. Before continuing, let me ask you the following questions. Do you tidy your house before visitors arrive? Do you go on a diet before going on holiday for fear of showing yourself? Do you enjoy other peoples' cooking and yet are highly critical of your own? After decorating your house do you notice every little blemish, paint dribble or flaw? Well, if so, it is probably because you, like so many people, can review your own work and are sometimes more critical than an independent eye. These case studies provide some evidence that the developer can effectively test their own systems.

There is a price for independent testing. A recent survey within Rolls-Royce (figure 1) showed that independent testing was costing up to 50% of the overall project costs - 25% on unit testing alone. If the relative cost for independence is so high, we must question if independent testing is actually a cost-effective process. If in some cases, independence is not effective, and funds are taken away from more effective and appropriate testing activities, this could be counter productive to product quality. However, if a more cost effective testing process were used, this would allow funds to be reinvested into the process to improve the quality of the testing.

6 Why Independence Does not Always Work

There are many reasons why independence does not always work, although it may be tempting to believe that with good management, sufficient documentation, good communication and skilled staff, many of these arguments do not apply. However, reality tends to pour cold water on theory. Many businesses are working under great pressure, staff are not always as experienced as required and projects never, ever, run smoothly. The following reasons are proposed for why independence alone does not always work:

6.1 Without a clear understanding of what you are testing against, testing is difficult if not impossible. Not all system knowledge can be easily explained and independent verifiers do not always have the right experience, background and the time to sufficiently understand the documentation.

6.2 Without system knowledge and a good understanding of the system, the independent verifier cannot evaluate the severity of the anomalies they find. Consequently, independent testing tends to find many minor anomalies of which most are not problems at all. This results in a large amount of unnecessary effort addressing these anomalies and may run the risk of drowning out a severe anomaly.

6.3 Without system knowledge, the independent verifiers tend only to find small local anomalies that can be logically understood in the limited context of the material under test. Wider issues and system anomalies may not always be spotted.

6.4 If you want to test something then you must develop the project with testing in mind. Independence removes the responsibility for designing for testability out of the hands of the developer. Developing a system to facilitate testing takes additional effort to implement and, unless there are strong motivational reasons, these issues are often overlooked.

6.5 Testing is often a post development activity. With the overheads of change management, anomalies tend to be ranked and all but the most serious justified rather then fixed. This is in part due to point 6.2 above, where many anomalies are not actually problems at all. Regardless, this can be demoralising for the independent verifier.

6.6 The independent verifier is not always aware where the risk areas of the system are and therefore may fail to adequately test them. Risk areas may include timing issues, stress or overload, interface and interrupt handling. Clearly, good risk management would help to reduce or alleviate this problem.

6.7 Just because testing is independent, it does not mean that an adequate amount of testing is performed. Unless testing is carried out by a separate organisation, there is still going to be a degree of organisational pressure. Even when testing is carried out by a separate organisation, (who submitted the lowest tender) there can still be contractual time-scale and financial pressures.

6.8 Do you really have independence anyway? If, to test something, you must first understand it, then is it possible that in understanding the system the independent verifier makes the same misunderstanding as the developer? Unless the independent verifier can think with an independent mind-set, it is unlikely that they will find errors different from those that a developer can.

6.9 A lecturer once described how when marking exam papers, students tended to get the same "difficult" questions wrong. In such a case, having students mark each others exam papers would not work. The verifier must be the most skilled person, whether they are developer or independent verifier.

7 Improving the Testing Process

The following six principles for improving the testing process are suggested:

7.1 Testers are not second class citizens. Put the testing in the hands of the most appropriate, skilled and experienced people whether they are developers or independent verifiers.

7.2 Decide how you are going to test a system at the beginning of the project and not at the end. Systems must be built around this requirement for testability.

7.3 To ensure that systems are more testable, you must put the responsibility of designing for testability firmly into the hands of the developers.

Designers must write tests in parallel with the development.

7.4 You must have something to test against. Ambiguous requirements are not testable. All requirements must be written clearly, precisely and in a form that can be tested.

7.5 You must understand what you are testing. The correct information and system knowledge must be available to the testers of the system.

7.6 Fix as many minor mistakes as possible before formally testing. Get it right before the code is frozen to reduce the costs of change control.

8 One Possible Solution

Given the problems identified in section 6 and the six steps for improvement proposed in section 7, the following life-cycle is proposed.

8.1 Define the requirements for independence and notify the independent verifiers. Their involvement throughout the project will ensure that they are not put under undue pressure at the end.

8.2 Develop the verification strategy for how the system will be verified. The strategy will cover documentation requirements, coverage metrics, review mechanisms, structure of system etc.

8.3 Submit the verification strategy to the independent verifiers for review. Any changes should be agreed and implemented.

8.4 For each requirement, the requirement authors shall write a set of tests. This will ensure that the authors consider testability from the outset and generate more testable requirements.

8.5 The requirements and tests shall be submitted to the independent verifiers for review. The independent verifiers shall identify additional tests if required.

8.6 The designers shall develop the high level design and integration tests and the low level design and unit tests. Again, this is likely to create a more testable system.

8.7 The design, integration tests and unit tests shall be submitted to the independent verifiers for review. The independent verifiers shall identify additional tests if required.

8.8 The coder will write the code and perform the unit tests. If static analysis is required, this may also be performed. Minor anomalies can be addressed immediately before the code is frozen.

8.9 The completed code shall be passed to the designers who shall perform, or take responsibility for, their integration tests. If necessary, the coders can perform the integration tests.

8.10 The completed code shall be passed to the requirement authors who shall perform, or take responsibility for, their system tests. If necessary, the coder can perform the system tests.

8.11 The test specifications, test reports and static analysis output shall be passed to the independent verifiers and reviewed. Deficiencies shall be reflected back to the developers.

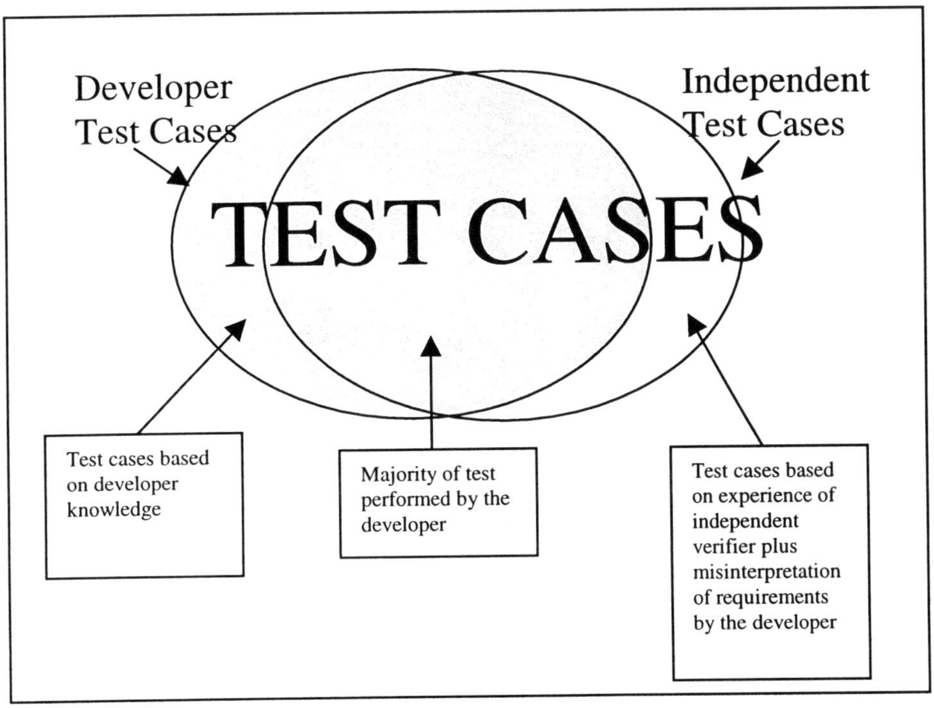

Figure 2: Test case generation by developer and independent verifier

The roles of requirement author, designer and coder need not be separate developers as implied by the life-cycle above. If necessary, all development roles may be performed by a single developer. However, by keeping these roles separate, this will introduce an additional degree of independence. As proposed, the majority of the test case and test executions will be performed by the developers (see figure 2).

The role of the independent verifier has moved more towards a role of reviewer rather then tester. The independent verifier would study the system and determine the requirements for testing. They will then consider the tests proposed by the developer. If there are deficiencies in the tests specified then these will be flagged to the developers for agreement and inclusion.

9 The Perceived Benefits of Do-It-Yourself Testing

The perceived benefits of using a more appropriate balance during testing between developer and independent verifier are listed below:

9.1 It is likely that no-one has a better understanding of a system than the original developer. The system knowledge that the original developer has is crucial to writing effective and rigorous tests and can compensate for any lack of understanding by the independent verifiers.

9.2 Because of the good understanding of the system, the developer can more effective judge the severity of any anomalies, thus reducing the number of incorrect anomalies identified. This will reduce bureaucracy if the total number of anomalies found is reduced.

9.3 You do not know if you can write a test for something until you try. If you cannot write a test for a requirement, it is probably not a requirement. By writing requirements that are testable, it is suggested that this will generate better requirements for both development and testing.

9.4 Requirement authors, designers and coders will develop more testable systems because they are forced to consider testability during development.

9.5 There will be a further reduction in project costs because the formal testing process will benefit from the informal tests that many developers perform as part of development.

9.6 The developers are more familiar with the systems, platforms, operating systems, tools and test environments then the independent verifier. With only the developers performing the tests, this will avoid the need for two learning curves as for when the independent verifier performs the tests.

9.7 If testing is performed during development, more of the minor anomalies are likely to be addressed than when testing is a post design activity and correction is a more expensive process. In some cases at SEAS, data flow, control flow and information flow static analysis is performed as part of the compilation activity, and all anomalies removed before the first line of code is frozen and executed.

9.8 Finding errors at the time the code is written will benefit from the design being fresh in the mind of the developer. Wait 3 months and even the original developer has to go through a learning curve to understand their own code.

9.9 Not only should the testing process be quicker but also, your own developers are contractually cheaper than those of an independent organisation.

9.10 The total amount of work for the independent verifier will reduce because they no longer need to write tests. Instead, they can focus their effort on studying and understanding the system. If the saving of effort, due to the developers performing the tests, is reinvested back into independent verification, this will increase the quality and benefit of the verification process.

10 Conclusions

There is not a single rule yet invented which satisfies all situations - the need for independence is no exception. Independence has been specified in the standards to resolve many of the problems identified by Myers. However, the current practice of independent testing is introducing other problems.

If there is even a single situation where independent testing is counter productive, we must reopen the debate and determine when independence is most appropriate. To find even a single situation where independence has been shown to be ineffective highlights a huge risk for those projects that blindly accept independent testing without question. We should not be using processes or methodologies on safety critical software if we are not fully aware of their purpose and benefit. The key to success is appropriateness, and appropriateness is only achieved through understanding. Do we really understand the benefits and costs of independence?

Although many organisations may perform their own testing activities either overtly or covertly, the standards do not reflect this practice. Not all organisations will be able to perform their own testing. Through a lack of resource or skill, many organisations may be forced to consider independence. This is not bad news but reflects the single issue of appropriateness. If you are unable to test effectively, then it is inappropriate for you to try.

Testing is one of the most important activities in the development process and should be put in the hands of the most skilled people whether they are developers or independent verifiers. The developer and independent verifier can bring different benefits to the testing process and it will be the pooling of resources between independence and dependence that will bring a higher quality of testing, at a lower cost.

Further investigation is required to determine an appropriate balance between the involvement of the developer and independent verifier. If further investigation suggests that there is a real benefit in changing the rules for independence then the standard authorities will need to be convinced for changes in current practice.

References

[00-55 1997] DEF STAN 00-55(Part 2)/issue 2 : Requirements for Safety Related Software in Defence Equipment: Part 2 : Guidance. Published by the Directorate of Standardisation: Ministry of Defence

[DO-178B 1992] RTCA/DO-178B : Software Considerations in Airborne Systems and Equipment Certification. Published by RTCA. Inc. 1140 Connecticut Avenue, N.W. Suite 1020 Washington DC

[EN-50128 1996] Draft European Standard prEN 50128 : Railway Applications - Software for Railway Control and Protection Systems. Published by BSI, 389 Chiswick High Road, London, W4 4AL.

[IEC-61508 1998] Draft International Standard IEC 61508-3: Functional Safety : Safety Related Systems part 3: Software Requirements (65A/181/CDV). Published by International Electrotechnical Commission

[Myers 1979] Glenford J Myers : Polytechnic Institute of New York The Art of Software Testing. Published by Wiley-Interscience, John Wiley & Sons

Can Formal Argumentation Raise our Confidence in Safe Design ?

David Robertson
Division of Informatics, University of Edinburgh
Edinburgh, Scotland

Abstract

It is technically possible to build systems of formal argumentation which help assemble evidence relating system designs to contextual material, such as safety guidelines. However, a number of assumptions underly this choice of architecture and influence its ability to support safer design. Many of these assumptions are not purely technical and apply regardless of the choice of formal representation. Using as an example a prototype argumentation system from a safety-related domain, a set of assumptions is identified and generalised to this class of system.

1 Background

The "Communication in Safety Cases" project[1] was a wide-ranging study of the technical, social and organisational factors which raise our confidence that we have designed safe systems. Part of the study looked at the potential to use formal representations of parts of the content of textual guidelines and, with the benefit of this underlying formal structure, to connect guidelines more closely to the design of safety shutdown systems. To demonstrate that our ideas were technically feasible, we produced a demonstration system which we summarise in Section 2 and which is discussed at greater length in [2]. However, technical feasibility doesn't imply that this type of system supports safe design or that the system itself is designed safely. These are important questions because our demonstrator is an example of a class of systems based on formal argumentation - which is the use of precise knowledge representation and inference methods to support the informal process of argument about some topic (in this case the design of a type of shutdown system). Such systems are becoming common in research environments and are finding their way into application (see [3] for a survey). The purpose of this paper is to raise some of the safety issues presented by such tools, using our system as an example.

The introduction of knowledge-based systems into safety-related areas has been a contentious subject since the early days of "expert systems" and remains so today, despite the fact that many of the techniques available are based on formal methods. One reason for this (discussed at length in [1]) is that many of the safety issues stem from assumptions which cannot be shown to hold unless we consider the lifecycle of design within which the knowledge-based system is situated. For this reason, we concentrate on those (often

[1]Funded by EPSRC/DTI under the recent Safety Critical Systems programme.

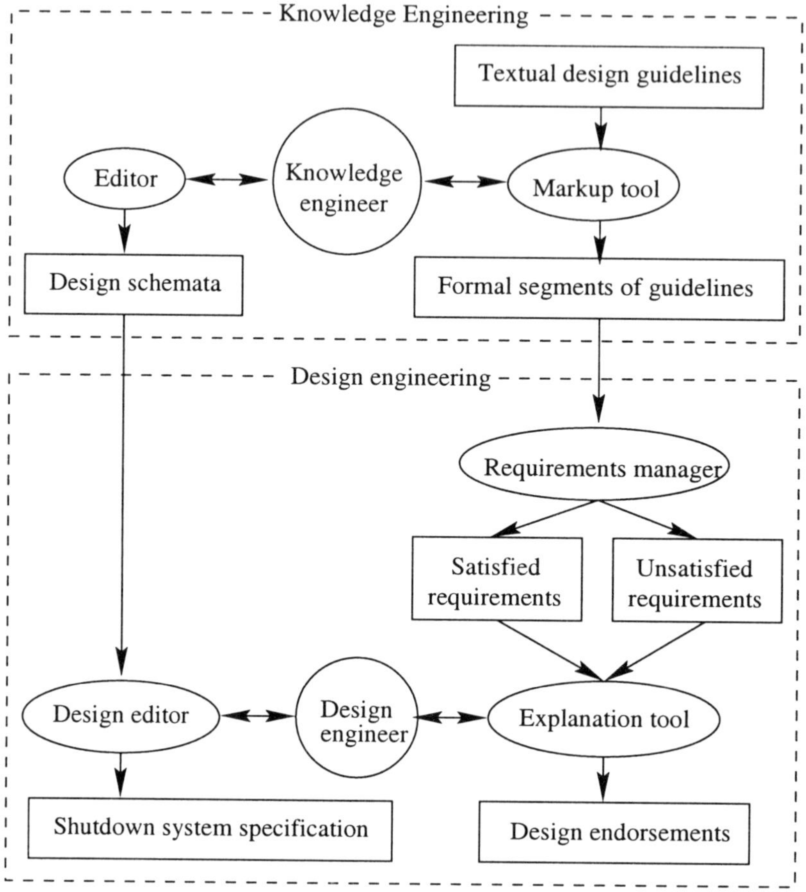

Figure 1: Overview of an argumentation system

neglected) contextual assumptions rather than on the more familiar technical aspects of the system (see [2] for these). In the next section we give enough technical information about the system to understand it from this point of view.

2 Structure of An Example Argumentation System

Figure 1 gives an overview of our demonstration system. It is divided into two interacting subsystems. The first, which we have called "knowledge engineering", concerns the formal representation of standard patterns of design ("schemata") in our target domain and the formalisation of segments of the textual design guidelines. The second, which we call "design engineering",

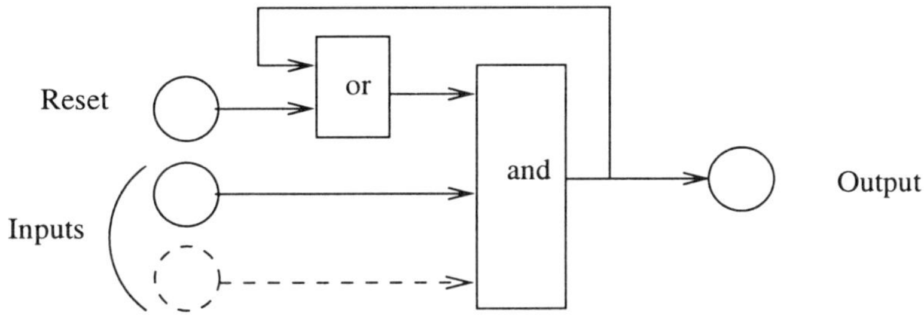

Figure 2: A reset schema

uses the knowledge represented in the first subsystem to help designers of shutdown specifications demonstrate that they are adhering to the design guidelines.

2.1 The Knowledge Engineering Subsystem

Figure 2 gives an example of a design pattern in our domain of application. This is a reset circuit connecting a reset button and one or more inputs to an appropriate output.

The function of this sort of circuit is to deactivate the output if any of the inputs are deactivated and to ensure that it is reactivated only if the reset and all the inputs are simultaneously active. The knowledge engineer must represent this in a formal style which can connect to other knowledge represented in the system. This is done using a formal structure containing three elements: a condition which relates the schema to structural features of the physical system to which the circuit will be attached; a model describing the behaviour of the class of circuits described by the schema; and a set of safety-related properties which circuits built using the schema should possess. A formal definition for this sort of schema, using predicate calculus, is given in Figure 3. The formal notation used is explained in Appendix A.

For those unfamiliar with predicate calculus, only one point is essential: the condition of the schema will bind the variables R, X, and $I_1, \cdots, I_n$ to a particular reset button, output and sequence of n inputs. This allows particular instances of the schema to be built by matching the condition the problem in hand. For example, if we know that $o1$ is an output which is locally reset by button $r1$ and the inputs expected to trip $o1$ are $i1$ and $i2$ then we can bind the condition of our schema as shown in Figure 4, with corresponding bindings in its model and safety properties. Those properties are that if either $i1$ or $i2$ are tripped this will cause shutdown of $o1$ and if $o1$ is shutdown and not reset then it will be latched (*i.e.* prevented from re-activating). We shall return to these properties later.

As well as describing design schemata, a knowledge engineer must also construct a formal representation of parts of the textual design guidelines which relate to the safety properties of schemata. For instance, one sentence

$$condition(output(X) \wedge local\_reset(R,X) \wedge \{I_1,\cdots,I_n\} = \{I|expected\_trip(I,X)\})$$

$$model(\left\{\begin{array}{lcl} activation(X,T,V) & \leftarrow & previous(T,T_p)\wedge \\ & & activation(X,T_p,V_0)\wedge \\ & & activation(R,T_p,V_r)\wedge \\ & & or\_gate(\{V_0,V_r\},V_p)\wedge \\ & & \left(\begin{array}{c} activation(I_1,T_p,V_1)\wedge \\ \vdots \\ activation(I_n,T_p,V_n) \end{array}\right)\wedge \\ & & and\_gate(\{V_p,V_1,\cdots,V_n\},V), \\ activation(X,T_i,V_i) & \leftarrow & initial\_activation(X,T_i,V_i) \end{array}\right\})$$

$$properties(\left\{\begin{array}{c} I \in \{I_1,\cdots,I_n\} \rightarrow causes(tripped(I),shutdown(X)), \\ causes(shutdown(X)\wedge not\_reset(X),latched(X)) \end{array}\right\})$$

Figure 3: A formal representation corresponding to Figure 2

$$condition(output(o1) \wedge local\_reset(r1,o1) \wedge \{i1,i2\} = \{I|expected\_trip(I,o1)\})$$

$$model(\left\{\begin{array}{lcl} activation(o1,T,V) & \leftarrow & previous(T,T_p)\wedge \\ & & activation(o1,T_p,V_0)\wedge \\ & & activation(r1,T_p,V_r)\wedge \\ & & or\_gate(\{V_0,V_r\},V_p)\wedge \\ & & \left(\begin{array}{c} activation(i1,T_p,V_1)\wedge \\ activation(i2,T_p,V_2) \end{array}\right)\wedge \\ & & and\_gate(\{V_p,V_1,V_2\},V), \\ activation(o1,T_i,V_i) & \leftarrow & initial\_activation(o1,T_i,V_i) \end{array}\right\})$$

$$properties(\left\{\begin{array}{c} I \in \{i1,i2\} \rightarrow causes(tripped(I),shutdown(o1)), \\ causes(shutdown(o1)\wedge not\_reset(o1),latched(o1)) \end{array}\right\})$$

Figure 4: An instance of the schema given in Figure 3

of the textual guidelines we used says:

> "All trip demands cause shutdown of the appropriate outputs."

Given the definition of our schema in Figure 3, a direct way of representing this requirement is as the following implication:

$$\forall X, I.output(X) \wedge expected\_trip(I, X) \rightarrow causes(tripped(I), shutdown(X))$$

2.2 The Design Engineering Subsystem

The purpose of the design engineering subsystem of Figure 1 is to connect the evolving design of the shutdown system to appropriate segments of the textual guidelines, using the formal representations of Section 2.1 as connective structure. The design engineer selects schemata from a library and creates instances of them as described in the previous section. In our system this is done using a graphical tool (the "Design editor" of Figure 1) which presents the schemata using a diagrammatic notation similar to that of Figure 2. This avoids design engineers having to learn predicate calculus - instead manipulating the underlying formal representation through a more familiar graphical language. Each time a schema is applied to create new parts of the shutdown system design it not only adds the corresponding model but also the safety properties which may connect the model to the formalised segments of the guidelines. Whenever convenient, the design engineer can ask for an explanation of which of the formal requirements relevant to the current design have been satisfied and which have not. To produce this explanation it is necessary to test the formal requirements against either the model itself or (more commonly) against the properties accumulated as part of schema selection. This requires a proof procedure, which it is sufficient to demonstrate here by an example (further details are in [2]).

Suppose that our current design contains only the output, $o1$; inputs $i1$ and $i2$ which we know should trip the output; and a reset design connecting these which we built by instantiating our schema in the way shown in Figure 4. We also have the formal requirement stipulated at the end of the previous section. To prove that it holds we shall assume that we have a "closed world" consisting of only the objects in the model and, based on this, prove the implication in our model by demonstrating that there is no known case in which it doesn't hold. The proof is as follows:

- We know that $output(o1)$, $expected\_trip(i1, o1)$ and $expected\_trip(i2, o1)$ are true and we want to prove:

  $$\forall X, I.output(X) \wedge expected\_trip(I, X) \rightarrow causes(tripped(I), shutdown(X))$$

- This is logically equivalent to the following statement:

  $$\neg(\exists X, I.output(X) \wedge expected\_trip(I, X) \wedge \neg causes(tripped(I), shutdown(X)))$$

- Given our "closed world" assumption, we can prove the above expression by showing that there is no instance of X and I such that the expression inside the outermost negation holds. The only instances are:

$$output(o1) \wedge expected\_trip(i1, o1) \wedge \neg causes(tripped(i1), shutdown(o1))$$
$$output(o1) \wedge expected\_trip(i2, o1) \wedge \neg causes(tripped(i2), shutdown(o1))$$

- Each of these is contradicted by the first property of the schema instance of Figure 4, which allows us to conclude $causes(tripped(i1), shutdown(o1))$ and $causes(tripped(i2), shutdown(o1)$.

Proofs like the one above connect parts of the design (in our example, the reset model of Figure 4) to the textual guidelines via the formalised requirements. This allows the system to do three things for the design engineer based on its attempt to prove any given formal requirement:

- If it is satisfied, it can describe the proof, so that the designer can check whether it is plausible.
- If it is not satisfied, it can explain where its attempt to find a proof failed, perhaps giving the designer a hint as to where the design may not fit the formal requirements.
- It can display the segment of text corresponding to the formal requirement in the guideline document, so the designer is alerted not only to the formal requirement but to its original textual expression and the context in which it was stated in that document.

This concludes our technical overview of the system. In the sections which follow we shall discuss some of the safety-related assumptions which underly this choice of architecture. The first set of these (Section 3) are assumptions about the structure of the system itself. The second set (Section 4) are assumptions influencing the contribution it makes to confidence that the design is safe.

3 Safety Issues Related to the Architecture

Our system relies on formal proof to relate guidelines to system designs so, at first sight, one might expect that safety arguments relating to the architecture of the system would revolve around questions of the correctness of its proof procedures. Although these are important, we argue in this section that other assumptions, which are indirectly related to the mechanics of proof, are of equal weight.

3.1 Appropriate System Modelling Assumptions

The schema given in Figure 3 contains a model expressed in a particular style, appropriate to a "ticking clock" view of the dynamics of the system. The idea is that there is a regular sequence at time points (referred to by the variable T in the model) and that change in the state of the system occurs by determining facts which are true at any given time point from those which were true at the previous time point (T_p in the model), with the recursive definition leading back to facts defined to be true at the initial time point (T_i in the model). This is a legitimate way of describing our target system but there are others which would allow us to model different aspects of the system - for example we might have chosen a modal temporal logic if we were interested in proving temporal properties from the model (*e.g.* that deactivation of any input will eventually deactivate the appropriate output).

Assumption 1 *If a system model is used to prove properties referenced by formal requirements then the style of modelling (at model level) must be appropriate to the form of proof required (at requirements level).*

The choice of modelling style (and of the logic to support it) only matters if we intend to perform proofs with it. If not, then all we need is some representation language which expresses the information found in engineering diagrams like the one in Figure 2 and has the appropriate facilities for parameterisation so that we can form instances of it (as in the transition from Figure 3 to Figure 4). It would be possible to use our system in this way, provided that the proofs connecting the formal requirements (from the textual guidelines) to the design could be done using only the properties given for each schema. This was, in fact, how we did the proof of the previous section. However, this is a weak use of the model itself, which becomes an "illustration" rather than being part of proof. If we use the model definitions as part of the proof then we rely on the assumption that the style of modelling permits the sorts of proofs which our formal requirements need, and that our proof procedure can find these quickly. Alternatively, if the model itself is not used in proof but the schema properties are, we have the issue of consistency between model and properties which we discuss below.

3.2 Property Verification Assumptions

The sets of properties associated with schemata are convenient ways of describing key features of parts of designs without having to prove these from "first principles" via the model itself. However, our system provides does not verify automatically that the properties we claim should hold for any instance of a schema actually will hold. For properties such as those in our example it is relatively easy to convince ourselves that there is no problem in this respect. For instance, the first of these says that tripping an input will cause shutdown of the output. If we interpret "shutdown" in our model as the variable V in

Figure 3 being bound to off in the predicate $activation(X, T, V)$) then we can see that changes in this value between time points occur only through the first clause of the model. Given the further assumption that 'and' gate sets its output to off if any of its inputs are set at off then it must be the case (via the first clause) that the output becomes off whenever this happens.

Assumption 2 *If schema properties, rather than a system model, are used to connect to formal requirements then we must assume that those properties hold for any model which can be created using the corresponding schema.*

Not all properties are so easy to demonstrate. For example, the notation used in our domain allows delay elements which allow the construction of circuits with behaviours which cannot be characterised simply by comparing adjacent time points. Proofs of properties for these could be more difficult and might, in general, be unavailable. If this were the case then our properties become, instead of guarantees, a description of the intention behind part of a design - the component may be intended to have a particular property (which may be relevant to the guidelines) but further proof and/or testing is needed to confirm that this intention is realised.

Assumption 3 *If the connection between schemata and requirements is through schema properties and assumption 2 doesn't hold then additional checking must be done to ensure the relevance of instances of properties to the model.*

3.3 Ontological Assumptions

As we explained in Section 2.1, a knowledge engineer has two tasks: building a library of schemata and constructing formal requirements from segments of the textual guidelines. When doing this, it is essential that the names used to describe different concepts are used uniformly in both schemata and formal requirements. To help establish this, a structure editor was provided for a later version of the system. This allows engineers to highlight a segment of the text and construct a formal requirement for it using a library of standard formal structures. Figure 5 shows a section of the guidelines with a segment highlighted and in the background the formal requirement editor with a menu of the available domain-specific structures. highlighted is the structure $carries\_gas(pipe(A))$ which denotes that A is a pipe carrying gas.

Figure 6 shows the editor in the foreground, after we have constructed a formal requirement. For convenience, it is divided into two parts: the goal expressed by the requirement and the condition under which the goal may be necessary. In this case the requirement state that if we have a single riser valve, C, located on some single riser then there should be an input device, A, located within the incident control centre and manual activation of A should cause the closure of valve C. This requirement was built by selecting general forms of expression from the menu and refining them with the assistance of background knowledge about the types of objects in this domain. For instance, the menu item highlighted in Figure 6

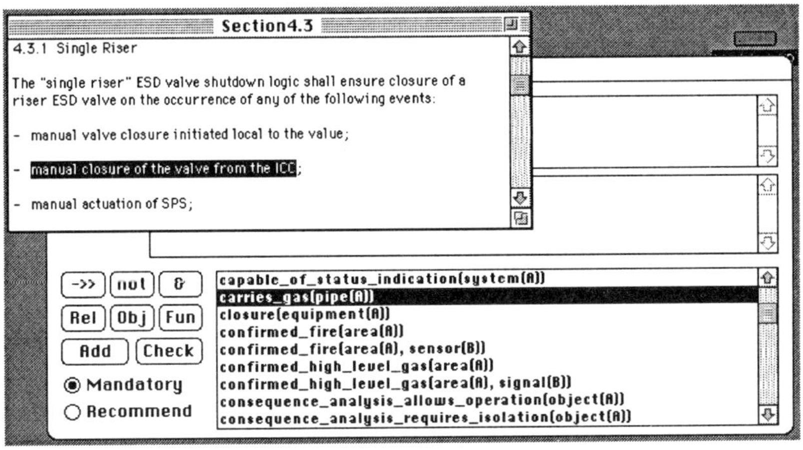

Figure 5: Highlighting a segment from the guidelines

$(located\_at(object(A), location(B)))$ can be refined to the first part of the requirement goal $(located\_at(input\_dev(A), within(icc(B))))$ because our background knowledge of objects gives us *input_dev* as a subtype of *object* and *within(icc)* as a subtype of *location*.

Although our structure editor can ensure that only formal expressions known from our library are used (so our formal language is "standard" in this narrow technical sense) it cannot ensure that these are interpreted uniformly by engineers in terms of the physical systems they wish to describe. What one engineer means by *non_essential* does not necessarily correspond to what another engineer might mean. This sort of consistency in description relies on human intuition for the appropriate formal "turn of phrase". To avoid re-invention of formal names for essentially the same concept, an engineer must be able to remember (or quickly search) the names which are available. This limits the size of the library of structures which we can offer.

Assumption 4 *The number of structures used in the ontology for formal requirements must be small enough that an engineer can recall them all during markup of text.*

Table 1 summarises the change in the size of this library when marking up a sequence of sections from the guidelines. The first column shows the section number (these done in ascending numerical order). The second column shows the number of formal requirements produced. These (like the numbers of sorts, functions and relations) are cumulative, so 60 requirements were described in Section 4.1; then a further 7 in Section 4.2 (giving a cumulative total of 67); then a further 42 in Section 4.3. The "sorts" column gives the number of sorts of object (names like *input_dev* in Figure 6) which we needed to define. This changed little between sections. Similarly the number of functions (names like *within* in Figure 6) was almost constant. The number of relations (names like

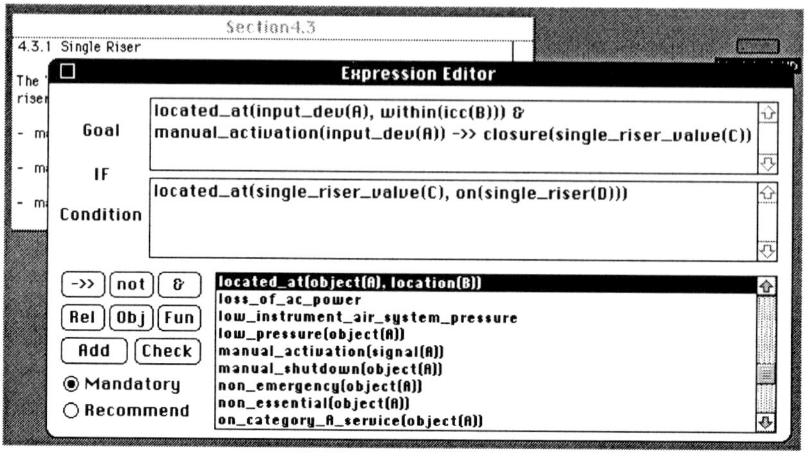

Figure 6: Structure editing a formula for a guideline segment

located_at in Figure 6) increased between sections but not as quickly as the number of formal requirements.

Section	Requirements	Sorts	Functions	Relations
Section 4.1	60	130	13	47
Section 4.2	67	130	13	52
Section 4.3	109	137	14	59

Table 1: Numbers of names used during markup

The information in Table 1 raises our confidence that a reasonably compact ontology for this domain might be obtainable, since the number of new names which we have to invent rises slowly as we add new formal requirements. However, we must interpret this result with care because one reason for the slow increase is that we chose to formalise those segments of the text which were easiest to do. The parts which seemed difficult were left aside. Therefore, we have a compact ontology but at the cost of reduced formal coverage of the textual guidelines, an issue which we consider in the next section.

Assumption 5 *Following from Assumption 4, if texts are large then it is likely that parts of them will not be given formal interpretations.*

3.4 Adequate Coverage Assumptions

As we saw in the previous section, the price of a compact ontology in this domain has been a selective formalisation of the text. Only those parts which are easy to interpret in our chosen style have been represented formally. We have therefore introduced the assumption that the coverage we have of the text is adequate for the library of schemata which we provide. Whether this is judged to hold depends on the purpose of the system. If its purpose is to guarantee that design engineers can be alerted to all requirements relevant to every schema then we must have a detailed argument which covers all of the textual guidelines. This would be difficult to do because what "relevance" means in this context is not determined purely mathematically - it can depend on what we think a given phrase is intended to mean. A more realistic aim is for partial coverage of those requirements which can be most directly interpreted in a formal way. This radically changes the purpose of the system from one of absolute guarantee to a "softer" form of support, using the formal structure as a route into the text but not using the formalisation as an alternative to the text.

Assumption 6 *Given Assumption 5, engineers must retain the textual guidelines, rather than their formalisation, as their ultimate point of reference.*

Figure 5 gives a concrete example of why the weaker assumption of "adequate" rather than full coverage is necessary. The text we have highlighted refers to "manual closure of the valve from the ICC', with the text above stipulating that the "The ... ESD valve shutdown logic shall ensure closure". However, our formal version of this (shown in Figure 6) doesn't stipulate which part of the system logic should cause closure of the valve - it only stipulates that the valve should close as a consequence of manual activation of an input device. Therefore, it is possible to argue that coverage of this textual requirement is not complete. Nevertheless, the formal requirement will be effective in detecting that the requirement is relevant whenever a single riser valve appears in the design and, if the requirement goal can be established automatically, it may also be helpful in suggesting whether the design is likely to be compliant. What it cannot do is to guarantee compliance. This must be done by referring to the original text, with the system being able to connect back to corresponding segments based on the markup.

4 Safety Issues related to the Application Environment

The discussion of the previous section introduced assumptions which accompany our choice of system architecture. A further set of assumptions must be made if we want to use the architecture to raise confidence in the safety of designs. These interact with the architectural assumptions we made earlier.

4.1 Proof Transparency Assumptions

The informal proof at the end of Section 2.2 is an example of the way in which schema properties (and hence parts of designs built using these schemata) are connected to formal requirements. Although based on one of the simpler styles of proof (see [2] for details) the full proof connecting requirements to schemata can be lengthy and complex. The explanation tool in our system provides conventional forms of support for engineers examining this proof: it allows the proof tree to be "walked through" in a structured way and paraphrases the formal expressions in stylised English. However, engineers who want to know exactly how a requirement was satisfied must understand the rudiments of this formal proof mechanism. We must assume that engineers have sufficient training to do this or, alternatively, that they use the proof system as a "black box" which flags potentially relevant requirements in the textual guidelines but does not reveal the underlying formal connections.

Assumption 7 *Engineers who need to understand the routes connecting models to formal requirements must have basic training in formal proof.*

Although the "black box" approach to proof asks less of design engineers it limits the help which the system can provide. In particular, for proofs which fail the system is able to give information about which goals it was unable to prove. An engineer with an understanding of formal proof might be able to use this information to help spot parts of the design which could be changed or improved. Were we to take this line then we must make a further assumption: that the goals which our system fails to prove are accurate predictors of weaknesses in the design. This isn't necessarily the case because it is possible that more than one way of proving as goal will exist and, if all the alternatives fail, it is not clear which should be blamed for the problem.

Assumption 8 *Engineers who do not need to understand how models and formal requirements connect also must not need to know the details of failure to satisfy requirements.*

4.2 Plausible Idealisation Assumptions

In Section 3.1 we raised assumptions about the appropriateness of system modelling with respect to the formal requirements. An equally important issue is the correspondence between our system model and the specification for the physical system. In our domain, there is a distinction between the logic of the shutdown system (which is what we call the "model" in this paper) and the layout of relays and other physical devices which will implement that logic (which is what we call the "specification" in this paper). One reason why this separation is useful in safety shutdown design is because the representation of the shutdown logic is more compact and, in some respects, clearer than the final system specification. However, it does this by presenting an idealised view of the system which has been proved useful in practice when producing specifications but which is untested when connecting to schema properties and thence to formal requirements. Only experience in use could tell us whether the idealisations we use in our models remain consistent when stretched this far across the design lifecycle.

Assumption 9 *We assume that the idealisations made to produce a compact model description make sense when used to discuss formal requirements and also when used to influence the design of specifications.*

5 Conclusion

All of the assumptions raised above are domain-specific so the answer to the question "Can Formal Argumentation Raise our Confidence in Safe Design?" cannot be answered from theory alone. We can, however, get closer to an answer by studying examples.

The system described in this paper is an example of an argumentation system, which uses formal methods for connecting system designs to contextual material (in our case design guidelines) relevant to arguments about the design. This architecture is unusual because it applies mechanisms of formal proof as a support for adhering to textual guidelines rather than as a formal replacement of those guidelines. Thus, it uses formal proof methods but those cannot be interpreted as a guarantee of compliance. The advantage of this approach is that it avoids an "all or nothing" approach to formality, in circumstances where "all" would probably be prohibitively expensive. However, its use relies on a number of assumptions which are not immediately obvious from the technical aspects of the architecture. The style of model and formal language used must be appropriate both to the demands of proof argumentation (Assumption 1) and to the domain (Assumption 9). If properties of the model, rather than the model itself are used to support proof then we must either show in advance that these properties always hold (Assumption 2) or build retrospective checking into our design process (Assumption 3). The ontology of our argumentation framework should be thoroughly understood by engineers so it must be compact (Assumption 4), which normally means that its coverage of the design context will be partial (Assumption 5). This, in turn, means that the informal contextual material cannot be thrown away when the formal structures are in place (Assumption 6) - they remain paramount. Although this can in some cases mean that design engineers need not know about the underlying formal system, since they are merely alerted to appropriate textual material via the automated system, if they ever want to discover how connections are made then some basic training in proof procedure is required (Assumption 7), particularly if they wish to learn from failed proofs (Assumption 8).

References

[1] J. Fox. Decision-support systems as safety-critical components: Towards a safety culture for medical informatics. *Methods of Information in Medicine*, 32:345–8, 1993.

[2] J. Hesketh, D. Robertson, N. Fuchs, and A. Bundy. Lightweight formalisation in support of requirements engineering. *Journal of Automated Software Engineering*, 5(2):183–210, 1998.

[3] P. Krause, J. Hesketh, and D. Robertson. Reliable and accountable system design. *Knowledge Engineering Review*, 12(3):289–305, 1997.

A Formal Notation

Term	Explanation in English
$A = B$	A and B are matching structures.
$C \leftarrow P$	C is true if P is true.
$A \wedge B$	A and B are both true.
$\{E_1, \cdots, E_n\}$	Set of n elements.
$\{X\|C\}$	The set of elements, X, for which C holds.
$activation(X, T, V)$	The activation at X at time T has value V.
$causes(A, B)$	If A holds then B will later hold as a consequence.
$condition(E)$	E is the condition of a schema.
$expected\_trip(I, X)$	It is expected that input I will trip output X.
$initial\_activation(X, T, V)$	The activation value for X at initial time T is V
$local\_reset(R, X)$	Button R is a local reset for output X.
$latched(X)$	X is maintained in a de-activated state.
$model(M)$	M defines the part of the system model contributed by a schema.
$not\_reset(X)$	X has not been re-activated as a consequence of reset.
$output(X)$	X is an output of the safety system.
$or\_gate(S, V)$	Sets V to 'on' if any of the inputs in set S are 'on', otherwise sets V to 'off'.
$properties(P)$	P defines the properties contributed to the design by a schema.
$previous(T, T_p)$	T_p is the time point immediately preceding T.
$shutdown(X)$	Component X is shut down.
$tripped(X)$	The element X is tripped (normally through de-activation).

Who can you trust? Assessing Professional Competence

R Malcolm, Ideo Ltd., London, UK
S Clarke, ERA Technology Ltd., Leatherhead, UK
S Hatton, ERA Technology Ltd., Leatherhead, UK
R May, CSE International, Flixborough, UK

Abstract

It is unarguable that those working on computer-based safety-related systems should be competent to perform their various tasks. Specifying what that competence should be, especially in a form which is assessable, has not been a simple task, but a set of competency statements and associated guidance on their interpretation and the operation of a competency assessment scheme is now available. Their development is described.

1 Introduction

In this paper is described the evolution over more than 12 years of the understanding, within the IEE and the BCS, of competence requirements for professional engineers working on programmable safety-related systems.

The culmination of this understanding will be the publication of a document containing a set of competence requirements for a range of safety-related activities, together with guidance on their interpretation and guidance on the way in which the competency statements may be used in a competence assessment scheme [IEE 1999].

2 The Stimulus

In 1986, a report to the U.K. Government Cabinet Office on the state of health of the U.K. IT industry had an appendix dedicated to the subject of safety-critical software [ACARD 1986]. That appendix led to a great deal of debate - and action

[Malcolm 1993]. In the context of the present paper, the key statements in that appendix were as follows:

> *The U.K. must develop the ability to certify safety aspects of software system construction and operation. These include:*
>
> ...
>
> * *certification of the software engineers who build and maintain the systems;*
>
> ...
>
> *Certification of software engineers will only be given when they have completed an approved level of full mathematical and methodological training together with an approved track record of experience.*

These statements reflected the emphasis at that time on software as software, rather than software as part of a system, and qualification through having taken particular accredited educational and training courses.

There were even proposals for the establishment of the position of 'Chartered Safety Critical Software Engineer', and a number of variations on this theme. Their common features were - typically - post-graduate training in particular techniques of software engineering.

3 The Response

The subject of individual certification was but one of the many areas we investigated in the IEE-BCS study which ran from 1987 to 1989 [IEE 1989]. In the proposal from the IEE to the Department of Trade and Industry for support for this study, we said on the subject of certification:

> *"During the last year the subject has been raised in the committees of both our professional bodies, and hotly debated in public meetings. It is a very contentious subject, with no consensus of opinion on the best way forward yet emerging. We therefore propose to canvass our members for their opinions of current proposals and elicit suggestions for other options for the future. After rationalising the responses we would again seek the views of the trade associations and other professional bodies on the technical, commercial, and legal consequences of these options."*

Two years later, the study concluded:

> *"We do not for now recommend certification, per se, in any form. ... We recommend that the safe operation of the most critical safety related systems should be the responsibility of a named individual within an accredited organisation. For any specific system there may be a number of such individuals within a variety of organisations responsible for different contributions to the overall system safety: an 'ultimate' individual - perhaps from a plant operating organisation or a system supplier - should be identified. Such named individuals should be 'competent'."*

4 Competence - a definition

Building on some of the ideas which came out of the IEE-BCS study, and drawing on the Ionising Radiation Regulations [HSC 1985], the IEE went on to develop a definition of competence. An initial version published in a 'professional brief' for members was later revised by the 'Inter-Institutional Group' of professional institutions and published in "The Hazards Forum Guidelines" [HF 1995].

First though, we established the requirement for competence. This was stated to be that *"All personnel dealing with safety-related systems (including purchasers, operators and maintainers as well as designers and implementors) should be competent to perform the tasks assigned to them."* (Later, a working group concerned with Railway Safety used the following variant: "All *personnel ... should be competent properly to exercise the authority delegated to them."* : an interesting distinction.)

We went on to give the following definition of competence:

> *Competence requires all practitioners to have qualifications, experience, and qualities appropriate to their duties. These include:*
>
> - *such training as would ensure acquisition of the necessary knowledge of the field for the tasks which they are required to perform*
> - *adequate knowledge of the hazards and failures of the equipment for which they are responsible*
> - *knowledge and understanding of the working practices used in the organisation for which they work*
> - *the ability to communicate effectively with their peers, with any staff working under their supervision, and with their supervisors*
> - *an appreciation of their own limitations and constraints, whether of knowledge, experience, facilities, resources, etc., and a willingness to point these out.*
>
> *Professionals with responsibility for design or for supervision of operators may be expected to have in addition:*
>
> - *a detailed working knowledge of all statutory provisions, approved codes of practice, other codes of practice, guidance material and other information relevant to their work; an awareness of legislation and practices, other than these, which might affect their work; and a general knowledge of working practices in other establishments of a similar type*
> - *an awareness of current developments in the field in which they work.*

This definition is inevitably very abstract, but it does capture the important point that competence involves much more than technical training, including attitude and behaviour as well as experience and knowledge of the application domain.

We felt it important to make this point strongly in order to counter calls for certification based solely upon qualification. Certainly, some qualifications are highly desirable, even necessary for some particular competencies. However, we were concerned that if we were to establish a scheme based solely on qualification, even as a stop-gap measure, it would be difficult to change its character in future. Moreover, such a scheme might be misinterpreted as a competence scheme, with potentially harmful effects.

However, while we may have captured the spirit of competence in our definition, to turn it into a set of more precise definitions appropriate for assessment would take a little longer.

5 Problems of Assessment of Competence

5.1 Ability and application

In the IEE-BCS Study we had already recognized that *"competence can never guarantee that any particular task is performed properly, since that depends on the correct application of that competence in the particular circumstances"*.

5.2 The past as a predictor of the future

We had also recognized that *"Competence depends not only on the capability of an individual, but on the appropriateness of that capability to the particular task in hand"*. If we are concerned with competence in wiring electrical plugs, then demonstration of capability to do this task previously means that one may presume competence to wire the next plug (apart from pathological problem cases). But a professional engineer is rarely, if ever, faced with performing tasks which are so similar.

This problem was recognized in "The Hazards Forum Guidelines". Here is another expression of the problem as it appeared in specific guidelines for the railway industry [not formally published] :

> *"Individual competence involves both knowledge and the ability to apply that knowledge. These are to greater or lesser degrees transferable from one work situation to another. At one extreme, such as making wiring connections in accordance with a wiring diagram, they may be very largely transferable. Such competence can readily be assessed and certificated. At the other extreme, the extent of the knowledge and ability which is transferable may be very limited. For example, an individual who may possess the knowledge and ability to develop computer software for engine management systems may not have the knowledge and ability needed to develop railway signalling software; and certification of competence in relation to engine management systems cannot be assumed to imply competence for working on signalling systems."*

So, having already said that it would be impossible to guarantee that someone would definitely perform a task 'correctly', we now moved a step further away, saying, in effect, that we could not either guarantee that someone could definitely perform a particular task. All we have is that to the best of somebody's judgment the individual should, in all probability, be able to do the task.

What we were, in effect, doing, although we did not articulate it in this way at the time, was distinguishing between the actual competence possessed by an individual, and competence as perceived, assessed, and possibly even 'certificated' by others. The former is what we really want: the latter is the best we can do with any competence assessment scheme. They are necessarily different.

As long as purists, including some of the authors, concentrated upon *actual* competence, we were unlikely to make progress in defining competencies in a way in which they could be assessed. The key to making progress was to shift from our original 'pure' definition of competence, which no-one could be sure of discerning, to a less strict definition, based on *perception* of competence on which we might find general agreement among those in our profession.

5.3 Objectivity, Subjectivity and Practicality

From the beginning, we have been concerned about the need, or at least the desire, for objectivity. In any competence assessment scheme participants will expect the assessment to be objective. However, there is a conflict between the objectivity of assessment and the value of that which is assessed.

It is easy to assess objectively whether someone has been on a training course or not: it is more difficult to determine whether they have assimilated what they were taught. During the early 90's this led us to think that we ought to examine engineers, rather than rely on their having participated on accredited courses. This idea was not pursued, and was overtaken, eventually, by the competencies study.

But the issue of objectivity continues to dog us. The judgment of whether someone has a background which means that they are able properly to perform some new task is essentially subjective. Since we want objectivity of a competence assessment scheme, should we give up?

No. Such theorising leads us to an impossible counsel of perfection.

Notwithstanding all the difficulties we have discussed, most systems are designed, built, and operated satisfactorily. In other words, most jobs are done competently by competent people. So someone, somewhere, somehow is satisfactorily making decisions about the competence of people to do these jobs. Our approach, then, has been to try to capture actual best practice, rather than define from a theoretical standpoint what we think should be done and call it 'best practice'.

6 The HSE Study

In 1994 the UK Health & Safety Executive and the IEE began negotiations which led to the HSE placing a contract with the IEE to work collaboratively with the BCS to develop a 'core set of competencies' in this area. This led in turn to a study, led by the IEE with ERA Technology as a major subcontractor, with CSE, Adval and McDermid Associates as subcontractors to ERA, and Ideo providing technical consultancy support to the IEE

A key requirement of the study was that it should be based upon a consensus of industry and the professional bodies, and to this end consultation was a major feature of the work. A Management Board, drawn from the IEE and the BCS was established, and a Steering Group with twenty or so senior representatives of different sectors of industry. In addition, there were targeted interviews with industry, a postal survey of industry, and widescale consultation within the IEE and the BCS.

6.1 Scope

The competencies considered here are restricted to safety-related activities. In some cases, such as Safe System Architectural Design, consideration of safety related systems concerns cannot sensibly be separated from other aspects of design, and the engineer would be expected to have appropriate competence in safety aspects of the job in addition to more general competence as a designer. In such cases the competencies are restricted to the 'safety value-added' aspects of the work.

This approach was taken to minimise overlap with and re-invention of competencies for activities which are not safety-related.

6.2 Genericity

It was recognized very early on that many aspects of competence depend upon knowledge of a particular engineering domain and experience gained in a particular application domain. However, our ambition has been to devise generic competencies both to facilitate movement of staff between domains and, importantly, to facilitate identification of mismatches between required competence and actual competence as people move to a new (to them) domain.

6.3 Comparisons with similar schemes

Similar work which has been carried out elsewhere was taken into account. In particular, the licensing scheme of the Institution of Railway Signal Engineers, the NVQ/SVQ competencies scheme, and the BCS Industry Structure Model were reviewed. These other schemes all helped to inform the design of the present framework. While we have stiven for compatibility, there are differences. These are generally a consequence of differences of purpose.

6.4 Importance of consultation

The extensive consultation with industry would have been desirable in any event, in order to avoid surprises and to gain the support of industry, rather than industry seeing the results foist upon them. These benefits have, we believe, for the most part been achieved. But capturing the best features of actual best practice has, in the opinion of the authors, been at least as valuable.

An additional benefit of the consultative approach is that we have been better able to achieve one of the other aims of the project, which is to try to ensure that any recommendations from the study should be compatible with existing competence assessment schemes in use in particular industrial organisations. The intention is to facilitate take-up of the results of this study, and to avoid additional costs for industry. Compatibility is a good aim, but given the diversity of existing competence assessment schemes, how could it be achieved?

6.5 Compatibility with existing in-house competency schemes

Although our intention was to design a competence assessment scheme which could fit alongside existing company schemes, we foresaw three possible areas of incompatibility.

The first problem was the potential for mismatch of structures. If an existing scheme has a simple structure then there is unlikely to be serious difficulty. However, if an existing scheme has a complex structure (though in general they do not) and the proposed new scheme were to have a complex structure, then there is much more likely to be a significant mismatch. And if the mapping between an existing in-house scheme and our proposed scheme were too complicated, then our scheme would be unlikely to be accepted.

We believe that this problem has been avoided by an approach in which the new scheme has a much simpler structure than first envisaged.

Given the diversity of schemes in use in different organisations, and the different motivation for the new scheme, there is unlikely to be a perfect fit with any existing scheme, let alone all schemes. Nevertheless, we believe that we have devised a scheme which can generally be mapped to most existing in-house schemes without great difficulty.

The other two concerns about incompatibility were whether the new scheme would go into too much or too little detail compared with existing schemes.

If the new scheme were not to consider competence in as much detail as present schemes, we would be adding nothing of value to the users of such schemes. So the new scheme would be unlikely to be adopted by those who, we presume, take the issue of competence most seriously. This would significantly jeopardize adoption of the scheme by them and, as a consequence, by others.

On the other hand, should the proposed scheme be - or appear to be - more complicated than an existing in-house scheme, adoption will again be jeopardized.

Eventually we settled on a structure which is much more simple than first proposed, in order to make it understandable to any potential user, and a level of detail which is toward the more sophisticated end of the spectrum of possibilities, but not as deep as the most sophisticated in-house schemes. To help adoption by those with much simpler existing schemes, and by those with no scheme at all, we have worked hard to keep both the structure and the presentation of the scheme as simple as possible, compatible with the chosen level of detail and the need to achieve, as far as possible, common interpretation by a wide range of readers.

7 Results of the Study - The Proposed Competency Framework

7.1 Task orientation

From the very beginning of our ruminations on competence we have emphasised that what matters is the ability to perform a task, rather than simply having the knowledge which, in theory, should enable the person to do the task.

What we have not yet discussed is the nature of a 'task'. How 'big' should a task be for the purpose of competence assessment? With how much detail should such a task be defined or described?

The first step was to decide how we should characterise tasks that people might do in terms of such things as industrial sector, accountability, application, regulatory regime, integrity level, and of course the degree of expertise brought to the task by the individual being assessed.

We considered a multi-dimensional grid to accommodate all or some of these dimensions, but realised early on that this approach would lead to far too complicated a structure. Not only would it be very difficult for potential users to understand; it would be near-impossible to maintain.

7.2 Functions

One route to simplification was to recognize that, whatever their environment, engineers usually contribute to recognizable functions, such as Safety-Related System Maintenance, and Safety Hazard and Risk Analysis. An individual may take total responsibility for a function (even, in principle, several) or may make a contribution to a function which, in total, is fulfilled by a team.

We identified twelve such functions in the first instance. These are:

- Corporate Functional Safety Management
- Project Safety Assurance Management
- Safety-Related System Maintenance
- Safety-Related System Procurement
- Independent Safety Assessment
- Safety Hazard and Risk Analysis
- Safety Requirements Specification
- Safety Validation
- Safety-Related System Architectural Design
- Safety-Related System Software Realisation
- Safety-Related System Hardware Realisation
- Human Factors Safety Engineering

Typically, to perform one of these function will require, in turn, performance of a set of more specific 'tasks'. For instance, we identified six tasks associated with the Safety Hazard and Risk Analysis function, two of which, for example, are 'Hazard identification' and 'Formation and control of hazard log'.

These tasks are the smallest elements of the proposed competency framework and the first step to defining actual competence requirements was to identify all the tasks associated with each function. Typically there are five or six such tasks. For each of these we defined a *'task-related competency'*.

Thus for the 'Hazard identification' task associated with the Safety Hazard and Risk Analysis function, the competency definition is:

Identifies hazards and hazardous events, including contributory and aggravating factors, for normal and degraded modes of operation through:

- *formation of a suitable hazard identification team (where appropriate)*
- *consideration of factors which could affect the operational environment and system performance*
- *use of appropriate historical information sources and hazard identification techniques.*

However, as indicated by our very early definition of competence, for a person to be considered competent, we expect them to possess certain qualities in addition to the technical ability to perform a task. So, for each function we have identified the key attributes required, and for each of them we defined competencies associated with such things as behaviour and understanding. Since these competencies are not specific to particular tasks of the function, they are known as *'function-related competencies'*.

While such competencies tend to be rather generic in nature, being concerned with such things as eliciting information and having personal integrity, their competency definitions are often tailored to suit the relevant function. Thus the 'team working' competence required for the same Safety Hazard and Risk Analysis function is defined as:

"Works well within a team and contributes effectively during hazard identification, analysis and risk assessment meetings/discussions."

For each of the twelve functions identified so far, we have defined up to 15 task-related or function-related competencies. Annex A is an example of a complete competency statement for the function from which extracts were quoted above - "Safety Hazard and Risk Analysis".

7.3 Levels

The set of competency statements for these twelve functions is the essential output of the study. However, we do not want just a yes/no decision as to whether someone is competent or not. We must recognize that different individuals bring differing degrees of expertise to their contribution to a function.

After considerable discussion and consultation, we now propose three 'levels' - expert, practitioner, and supervised practitioner. (We actually felt that Master, Journeyman and Apprentice better captured the spirit of differentiation and level of competence we sought, but were persuaded that such names would not be acceptable in the modern world.)

These levels will be described in some detail in the final publication, but for our present purposes they are near-enough self-explanatory. An expert is someone that most people would agree is an expert. A practitioner is someone who can be given significant responsibility for a function. A supervised practitioner is not a complete novice: they are expected to have sufficient understanding of best practice to be able to work *"without placing an excessive burden on the [supervising] Practitioner or Expert"* even though they may not have had previous experience of working on a safety-related project.

Three levels are not as many as may be found in some existing in-house schemes, but often such schemes have such a fine grading not for the purpose of matching competence to requirements for a job, but to provide a career progression 'ladder'. Although we must recognise that competence assessment against the proposed definitions will almost certainly be used to indicate career progression, we did not want this scheme to be explicitly linked to staff appraisal schemes, or primarily used for that purpose. We also doubt whether we could define competence guidance in a way which discriminates between more than three levels in a way which is generic and generally acceptable.

7.4 Context

So, we succeeded in generalising away from specific details of tasks. However, while this yields generic competencies, as we desired, we now have the need to determine whether an individual has competence in the specific required application domain, industry sector, integrity level, etc.

This information is captured in 'context information'. During a competency assessment for an individual, it is necessary to specify the context in which competency is required by the organisation and, where it is different, the context in which the individual established their competency.

8 Guidance on Interpretation of the Competency Statements

While the competency statements themselves are succinct and generic, they are also, thereby, quite abstract. We felt it necessary to give some indication of how those statements should be interpreted.

We have therefore developed additional guidance material to indicate, primarily, the kind of evidence required to demonstrate each type of competency at each of the three 'levels'.

Implicit here is the importance we attach to evidence. This is not because of a of lack of trust in the statements and opinions of engineers, but is necessary in order to provide traceability and auditability.

The nature of the evidence which might be provided can be characterised in two ways. First and foremost is the 'level' of evidence, alluding to the level of competence of the individual. For instance, a practitioner is likely to be expected to provide evidence of having performed the task before, whereas a supervised practitioner may be able to provide only evidence of having been trained to perform the task.

The second characterisation is in terms of the material form of the evidence. Several options are considered, including the preferred form of documentary evidence associated with previous work on the task, but also such things as interview reports and work place observation.

Annex B is an example of the guidance for the "Safety Hazard and Risk Analysis" function.

9 Trials

As part of the consultation process, the project Steering Group took part in two rounds of trials in which they 'role played' in assessments of actual members of the group, using the competency statements and assessment guidance emerging from the study. This was a particularly important validation exercise and, encouragingly, proved successful. That is to say that different members of the Steering Group, with very different backgrounds, agreed on the interpretation of the statements and the guidance, so that they agreed on the competency of the assessed members of the group.

In other words, the statements and the assessments based upon them, were 'sufficiently objective'.

10 Guidance on the Operation of a Competency Assessment Scheme

In addition to providing guidance on the interpretation of the competency statements, the competency publication will also contain guidance on the way in which a competency scheme for safety-related practitioners could be operated in a organisation. It will suggest an approach to allocation of responsibilities for management of the scheme; mapping of the scheme to the organisation's structure and practices; scheduling of competency assessments; and the actual performance of a specific competency assessment.

11 Related developments

11.1 IEC 61508

While many standards in the past have specified that people should be competent, as long as compliance with the requirement was only 'a good thing', non-compliance was not usually a big issue.

However the IEC 61508 standard says, for instance: *"assurance that the safety integrity target has been satisfied for systematic failures is gained by ... the use of competent staff"*.

In other words, it is fundamental to safety that the staff are competent, rather than just being 'nice to have'.

11.2 Safety Related Systems Certification

Most existing certification is for products, rather than complete systems. In parallel with the competencies study, a separate project (CASS) has been investigating the prospects for certification of whole systems and subsystems for conformance with IEC 61508.

It has become apparent that there is a difference in nature between certification of a general purpose product, such as a sensor - even a smart sensor, and certification of an installed fire and gas system for an oil platform. The former typically depends primarily upon conformance with a specification: the latter is heavily dependent on the competence of the engineer to configure and parameterise the system, based upon understanding of the domain and the equipment. In addition, there is even greater dependence in the second case on the management processes of the organisation.

The implication is that if there is to be such a conformance assessment scheme, it will necessarily lean on the assessment of competence of individuals. The publication of competency statements is therefore timely.

11.3 Licensing and Accreditation

Neither licensing of engineers, nor establishment of an accredited assessment scheme, have been addressed yet. However, these possibilities have been borne in mind during the definition of these competencies so that the approach adopted does not preclude the future introduction of licensing or an accredited scheme, should industry wish such a scheme to be established.

12 Summary

We have described the development of a set of competency statements and associated assessment guidance for safety-related systems practitioners. We have also explained some of the rationale for the proposed competency framework. A more detailed rationale which explains specific decisions about the design of the competency framework will be provided along with the competency statements.

The competency statements and the associated guidance on their interpretation and the operation of a competency assessment scheme are now available for comment prior to final publication [IEE 1999].

Acknowledgments

The authors would like to acknowledge the very many people who have contributed to the work described, including all the work within the institutions leading to the recent study. In particular, we thank the members of the Management Board of the study for their guidance and patience. The Board was chaired by Professor Andrew McGettrick of the University of Strathclyde, and included Stuart Gunn of the IEE, Arthur Lawrence of British Energy, David Newman, formerly of the Ford Motor Company, Malcolm Sillars of the BCS, and Ray Ward of the HSE. We also thank for their contributions the other members of the project team, these being Audrey Canning, formerly of ERA Technology and now of Virkonnen Ltd, Valerie Downes of AdVal, and Professor John McDermid of McDermid Associates.

References

[ACARD 1986] *Software: a vital key to UK competitiveness.,* HMSO, 1986.

[HF 1995] *Safety Related Systems - Guidance for Engineers Issue No. 1* (ISBN 0 9525 103 0 8) Hazards Forum 1995

[HSC 1985] Health and Safety Commission *Protection of persons against ionising radiation arising from any work activity:* HMSO, 1985

[IEE 1989] Institution of Electrical Engineers and the British Computer Society: *Software in safety-related systems: a report prepared by a joint project team* (isbn 0852963 91 2) IEE, 1989

[IEE 1999] *In safe hands: Competencies for Safety-Related System Practitioners* IEE 1999 (Note .. at the time of writing, details of publication had still to be finalised. Contact the Public Affairs Board of the IEE +44-171-240-1871 for further information.)

[Malcolm 1993] Malcolm, R E: *The JFIT Safety Critical Systems Research Programme: origins and intentions* In Redmill, F & Anderson, T: *Safety-Critical Systems - Current Issues, Techniques, and Standards* Chapman & Hall, 1993

ANNEX A: COMPETENCY STATEMENT EXAMPLE

Function	Safety Hazard and Risk Analysis	Version F

Summary

Safety Hazard and Risk Analysis involves responsibilities for identifying all foreseeable hazards and assessing the risk of an accident. Additionally there is a responsibility to ensure that the results of the hazard and risk analysis activities are documented and that the hazard log is controlled throughout the development (and often throughout the lifetime of the system).

In discharging these responsibilities the key tasks for the function are:
- identifying hazards associated with the process, the system and the environment, including defining the scope of the analysis
- analysing hazards for their impact on the safety-related system and assessing the level of risk
- assisting with the identification of mitigations to eliminate hazards or reduce the risk of accidents
- documenting the results of the hazard and risk analysis activities and managing the hazard log.

Task-Related Competencies

HRA1 Defining the scope of a hazard and risk analysis. Gains familiarity with the context and scope of a hazardous problem (process, equipment and personnel) sufficient to perform a hazard and risk analysis, and defines the scope clearly in a written document.

HRA2 Hazard identification. Identifies hazards and hazardous events, including contributory and aggravating factors, for normal and degraded modes of operation through:
- formation of a suitable hazard identification team (where appropriate)
- consideration of factors which could affect the operational environment and system performance
- use of appropriate historical information sources and hazard identification techniques.

HRA3 Hazard analysis. Analyses whether a hazard can be caused by the behaviour of the safety-related system and how a hazard can lead to an accident by:
- systematically deriving, collecting and analysing relevant information in determining hazardous system behaviours
- analysing the operation and maintenance aspects of the system
- determining and analysing accident sequences.

HRA4 Risk assessment. Determines the consequences and frequencies of accidents associated with the occurrence of hazardous events by selecting and applying an appropriate risk assessment method (e.g. the risk graph).

HRA5 Eliminating or mitigating hazards. Identifies potential risk reduction measures and evaluates them using ALARP techniques, including consideration of:

- the use of different technology/solutions
- the wider environment including external risk reduction features such as other systems, devices and operation and maintenance activities
- the relevant safety regulatory requirements, legal issues and the tolerability of risks within the relevant sector.

HRA6 Formation and control of hazard log. Produces a hazard log documenting the results of the hazard analysis and risk assessment activities which can be used to monitor and control the hazards throughout the lifetime of the system.

Function-Related Competencies

HRA7 Principles of functional safety assurance. Addresses the principles of functional safety assurance especially hazards, risks, tolerability ALARP and SILs in all hazard and risk analysis activities and understands how the hazard analysis and risk assessment activities impact a safety case.

HRA8 Application domain knowledge. Considers the process, external equipment, the operating environment, maintenance activities and other human interactions, etc. associated with a safety-related system throughout the performance of hazard and risk analysis activities.

HRA9 Systematic approach. Employs systematic methods of identifying, analysing and assessing hazards to ensure that all aspects of the behaviour of the system in its environment are addressed.

HRA10 Systems viewpoint. Considers the overall system within the operation, process and the wider environment and is able to abstract away from unimportant detail.

HRA11 Professional standing. Has a level of professional standing sufficient to give credibility to judgements on hazards, risk assessment and safety engineering issues.

HRA12 Team working. Works well within a team and contributes effectively during hazard identification, analysis and risk assessment meetings/discussions.

ANNEX B: Assessment Guidance Examples

Function	Safety Hazard and Risk Analysis	Version F

Summary

Safety Hazard and Risk Analysis involves responsibilities for identifying all foreseeable hazards and assessing the risk of an accident. Additionally there is a responsibility to ensure that the results of the hazard and risk analysis activities are documented and that the hazard log is controlled throughout the development.

In discharging these responsibilities the key tasks for the function are:

- identifying hazards associated with the process, the system and the environment, including defining the scope of the analysis
- analysing hazards for their impact on the safety-related system and assessing the level of risk
- assisting with the identification of mitigations to eliminate hazards or reduce the risk of accidents
- documenting the results of the hazard and risk analysis activities and managing the hazard log.

Task-Related Competencies

(NB – just one of six which is shown here as an illustration)

<table>
<tr><td colspan="3">HRA2 Identifying hazards</td></tr>
<tr><td colspan="3">Identifies hazards and hazardous events, including contributory and aggravating factors, for normal and degraded modes of operation through:
• formation of a suitable hazard identification team (where appropriate)
• consideration of factors which could affect the operational environment and system performance
• use of appropriate historical information sources and hazard identification techniques.</td></tr>
<tr><th>Supervised Practitioner</th><th>Practitioner</th><th>Expert</th></tr>
<tr><td>For a given scenario, can identify hazards.</td><td>Explains the difference between hazards, accidents and failures.</td><td>Understands how definition of a system boundary can affect the definition of hazards and failures and the relationship of a system to external events and accidents.</td></tr>
</table>

HRA2 Identifying hazards		
Given a particular safety-related system scenario, with which he/she is reasonably familiar, displays lateral thinking ability in identifying associated hazardous situations.	*Presents evidence of the performance of hazard identification exercises (e.g. brainstorms) that demonstrate the use of lateral thinking in identifying hazards and, in particular, hazards relating to the operation and maintenance of the system and degraded modes of operation. Shows how previous designs, incidents and other information sources have been used in hazard identification exercises.*	*Illustrates with documentary evidence (e.g. hazard identification procedures and training course notes) how hazard identification teams are selected such that an adequate degree of lateral thinking ability is present within the team, and how team activities are run to ensure sufficient results are obtained within a reasonable amount of time.*

Function-Related Competencies

(NB – just one of six which is shown here as an illustration)

<table>
<tr><th colspan="3">HRA12 Team working</th></tr>
<tr><td colspan="3">Works well within a team and contributes effectively during hazard identification, analysis and risk assessment meetings/discussions.</td></tr>
<tr><th>Supervised Practitioner</th><th>Practitioner</th><th>Expert</th></tr>
<tr><td colspan="2">Works as an effective member of a hazard and risk assessment team. Shows how he/she has contributed to hazard and risk analysis work as part of a team.</td><td>Identifies key attributes of a hazard analysis and risk assessment team. Describes situations where team dynamics has led to incomplete or insufficient analysis.</td></tr>
</table>

AUTHOR INDEX

Towards System Safety contains the invited papers presented at the seventh annual Safety-critical Systems Symposium, held in Huntingdon, February 1999.

The papers bring together practitioners and researchers in a quest to inculate a higher degree of safety engineering into the development and operation of critical software-based systems. In particular, this volume explores recent work and experience which leads us further 'towards system safety'.

ISBN

.co.uk